GOLD REFINING

2nd edition

by GEORGE J. GAJDA

Published by the author

P.O. Box 1846, Santa Monica, California 90406.

ISBN # 0-9608018-2-0

Printed in the United States of America.

TABLE OF CONTENTS

"The author wishes to acknowledge the help of Alice Mohr for editing and arranging the contents of this book."

The author (left) with the chemist at HARLYN PRODUCT Inc.
Los Angeles, California.

METHODS OF GOLD RECOVERY

Introduction:

The primary source of gold metal is mining operations, directed toward production of precious metals. But appreciable quantities of precious metals are found associated with other base metals and recovered incidentally during the refining of these base metals as a by-product. In recent years the recovery of secondary gold originating from other sources than mining operations, exceeded the volume of gold metal from primary sources. This is correlated to increase in application of gold in industrial products and consumption of gold by industry.

The scope of this book will handle the methods of recoveries and refining from primary and secondary sources of precious metals such as major industrial uses in jewelry, electronics and dentistry. Until 1969 gold processing was regulated by the Federal Department of Treasury. Gold for industry was available for purchase only from the U.S. Mint. Since 1969, the purchases of gold from industrial applications were not from the mint, but directly

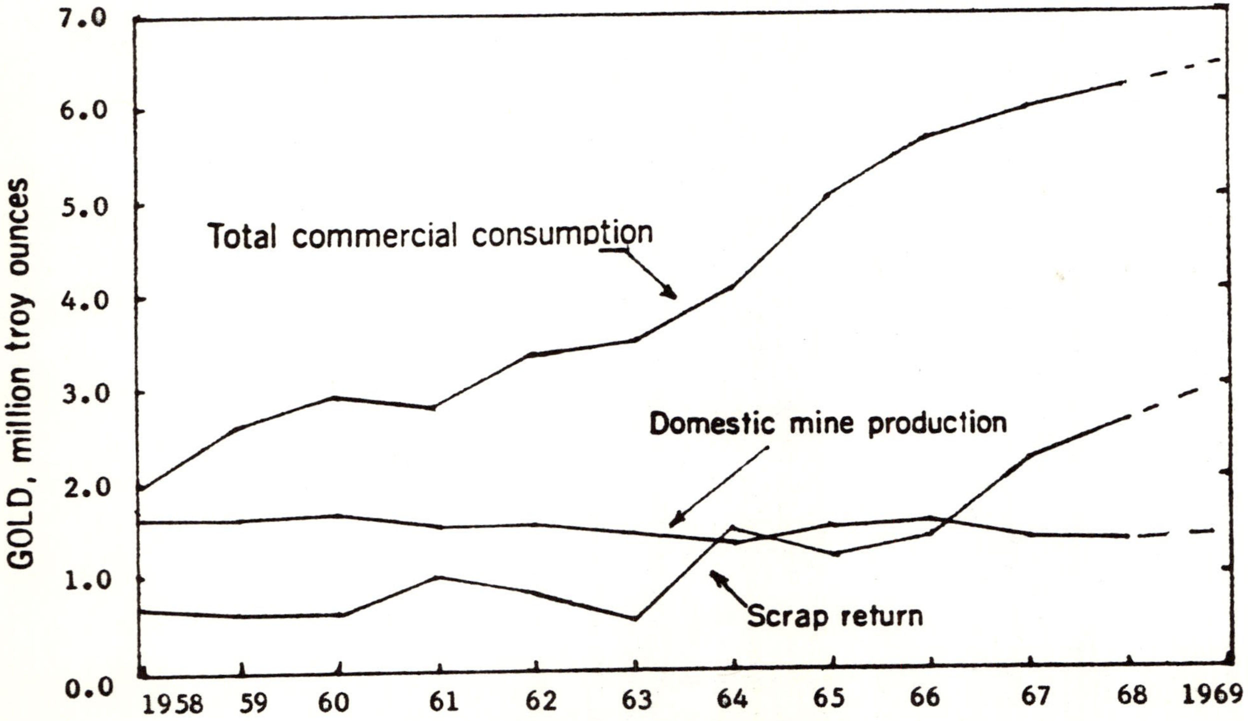

Consumption and Sources of Gold.

from the gold refiners, who in turn were controlled by the Bureau Of Domestic Gold And Silver Operations of the Department of the Treasury.

There were only a limited number of licences issued up till 1974, not in excess of 1200 for the whole United States. From January 1974 the licensing and limitations for gold processing were lifted and gold became a free commodity and, therefore, anyone since then, can engage in refining and industrial processing of gold. This may be the reason why there is an obvious lack of information concerning these operations, which were confined to a small segment of the industry. At least very few references, papers or books, can be found on processing and refining of precious metals.

The free market of gold in the United States was expected to produce a new value and price of precious metals, namely gold. However, surprisingly small and rather finite market of gold is still ruled by the old market interests in London, Paris, and Switzerland, which are fixing the price of gold, daily, according to supply and demand for the entire world production.

In the United States, some efforts have been made in regulating the market price. American Express Company entered the gold marketing field in 1967 with a license of 75,000 Troy ounces, hoping to establish a New York market for gold. They discontinued this activity in 1973. At the present time some indications are that the gold market price for the United States is established by the Chicago industry. For most of the cases, the gold prices are indicated by Englehart Co. or Handy and Harmon Co. which generally run some small percent of the price over the London second fixing, and thus announcing that at that price, all the gold they need for their own industrial purposes is available.

The purpose of the book is to inform the reader on the general market of precious metals in the United States, the practices in the industry and the methods by which the gold can be re-introduced in a marketable form.

1. The first consideration has to given to the market itself whether the gold will be in a semi-processed form, eg; bullion, for the uses of Industry or if the gold can be fabricated into salable items.

2. Second consideration is given to the total technical involvement and the extent of work necessary for each and particular instance of gold processing.

The gold which is purchased for processing and refining can originate from any conceivable source, in any type of form or quality. The knowledge of the material, the capability of recognizing the true value of the gold bearing material is probably the basic necessity in all cases.

It is in the interest of the reader who has neither any previous experiences in the field of industrial refining of precious metals nor has knowledge of chemistry, to introduce the basic information on the subject, which usually escape the attention of writers, concerned about the extent and variety of processes in the industry. The knowledge of the materials and their chemical behavior surpasses any other experiences in the field of precious metals refining.

It just happens that the great portion of the gold traffic in the United States and for this matter, in the whole world, is in small businesses of jewelry manufacturing, practicing dentists and gold scrap re-sellers. The multitude of small businesses including dental laboratories and dentists are simply not equipped to recognize the quality or purity of gold. Typical example can be given on application of dental alloys by professionals that simply take for granted the type of alloy they receive from the supplier and satisfy themselves with the description of the alloy, percentage of the gold, and physical qualities of the alloy. This information however, is not forwarded to the patient or buyer of the dental scrap and in some instances it cannot even be traced back to the origin of the material. Consequently, the prices offered for dental scrap are guess work and only by experience with repeated sources can such an evaluation be sufficiently accurate. The gold originated from jewelry is also very difficult to evaluate and it is usually done by identification stamping by the manufacturer. Jewelry stamped by number of Karats does not necessarily have any relationship to the percentage of gold in the total weight of the jewelry, due to the fact that some settings of the stones require different alloy and the volumes weight of the stones render the determination of the specific density impossible. Moreover such stamps are available to any one, and even to people who may abuse the buyer. The purchase of gold which originates from mines and small operations presents also a great risk to the buyer. For these reasons it becomes evident that a thorough examination of the material is necessary and also to determine chemically the percentage of gold in gold bearing materials at least to the accuracy which would not allow a significant loss in monetary value after the purchase is closed. This leads to the conclusion that the most important part of precious metals recovery and refining is the capability to determine the percentage of gold in the material prior to the processing. Once the processing of the batch of the material has started and has been carried to the final stage, the expected value of the material may be past redeeming. The losses which occur may be aggravating to both parties, the processor who lost the money, and the seller, who may never be sure that he received a fair price. It is not practical for a small businessman or scrap dealer to have a chemical laboratory or expensive instruments for determinig the chemical qualities of the precious metals. It is also impractical to use in every case, at the occasion of purchase, the services of a commercial laboratory which may charge substantial fees for determining the percentage of gold on selected samples. The analytical results can certify only the chemical composition of the sample which was submitted. This analysis may not have any relationship to the bulk of the purchased material.

It may be concluded that a well equipped laboratory will be an important asset to the refiner and within its capability, it can also assume a function of a small batch refining facility. As in many other instances the equipment of the laboratory could vary from simple to very complicated instrumentation depending on the volume of work, the initial investment, and the capability of the personnel.

The very basic instruments in laboratories will consist of analytical balances and scales to determine the weight of the materials. There should be under any conditions at least three kinds of scales. First to determine the weight between zero and 200 U.S. pounds with the accuracy of 0.25 lb. Second the intermediate range of metric balances between zero and 2,500 grams with accuracy of 0.1 gram, and the analytical macro balances with a range of zero to 200 grams with the accuracy of 0.1 of a milligram. At this particular point it should be emphasized that the metric system is essential in any chemical work. The reader should get acquainted with the metric system at least to the extent of converting the volumes and weights to a common denominator which in this case is a gram or cubic centimeter (a milliliter).

The market price of gold is quoted per Troy ounce. A Troy ounce of gold is equal to 31.10348 grams. The factor of 31.1 is sufficient for any calculation. It is very important to realize that there is no relationship between U.S. ounces or pounds to Troy ounces or Troy pounds used to express weight of precious metals. In practice, however, materials may be weighed in U.S. ounces or pounds. In order to convert these values from pounds to Troy ounces, we shall multiply the pounds by 453.6 and divide by 31.1. We find that a U.S. pound has 14.58 Troy ounces. A Troy pound has 12 Troy ounces which means 373.24 grams and its weight is only 82% of the U.S. Pound. It is obvious that it is very important to understand the relationship between the weights and other units of the existing system and therefore it is recommended to use preferably the metric system. After obtaining the results using metric units, we may convert these units to any commonly used units which are negotiable. For limited use, the tables of conversion factors printed on page 56 are sufficient.

CHAPTER ONE

THE CHEMICAL PROPERTIES OF GOLD

Chemically gold is an inactive metal. It is not attacked by oxygen, and therefore, it is relatively corrosion free. In massive form it is practically insoluble in any single acid, however, gold is attacked readily by halogens in presence of oxygen, and with the exception of fluorine, it combines to form auric chlorides at temperatures of 150o C. It also combines slowly with bromine and iodine. In presence of strong oxidizing agents, it will be attacked by sulphuric acid, and very actively by hydrochloric acid. The aqua regia, is the strongest solvent for gold. The aqua regia is a combination of hydrochloric acid and a strong oxidizing agent such as nitric acid. It will readily dissolve gold into oxychlorides and aurichlorides. Gold also, when heated with alkalies such as KOH is attacked and converted into soluble alkali aurates (KAuO2).

Gold forms many complex compounds. The common valence of gold is positive three. It also displays a common valence of positive one, but the aurous compounds tend to change spontaneously into auric compounds and metallic gold when heated in solution. Both, aurous and auric chlorides, dissolve in excess of hydrochloric acid.

The gold readily dissolves in sodium or potassium salts of cyanide in the presence of oxygen. This reaction is very usable for processing of gold. The gold can be quantitatively recovered and therefore, it is the basis of the recovery of gold from ores and other sources.

The insolubility of gold in common acids is sometimes used as means of identification or differentiation of gold and other yellow alloys of copper. This, however, is not a fool-proof method It may be indicative but not conclusive. In any case, the insolubility of materials cannot conclusively determine the presence of gold. Therefore, The value of the material remains unknown. It is very desirable to evaluate materials by qualitative and quantitative methods in order to determine the purchasing price of the materials. In next chapters the methods-which are conclusive in quantitative determination of elements will be discussed and reference will be made to standard analytical methods.

Gold usually is associated with metals of its own group (family), the copper and silver. These three elements display a pronounced tendency to form very stable complexes. Copper, silver and gold are not readily oxidized and are found on the bottom of electro-chemical series while the alkaline metals are at the top. These three metals are basically the major constituents of gold alloys found in jewelry or dentistry. A percentage combination of each is producing a yellow color of an alloy which is indistinguishable from pure gold, however, the physical properties of these alloys may substantially be different from each other.

It is an old accepted way of describing the quality of gold by indicating the number of parts of gold in 24 parts of the metal. Therefore, a 14K yellow Gold contains 14 parts of gold in 24 parts of the metal or, 58% of gold (100% ÷ 24 - 4.16% x 14 - 58.3%.

This alloy also contains approximately 25% of copper and silver 15%. The remaining 1.7% contain other minor constituents which are added usually to improve the hardness and grain structure of the alloy. Similarly, 18K gold for most of the time contains 75% of pure gold with a minimum of 18% copper and remaining percentage of silver. The color of gold is strongly affected by the relative proportion of copper to silver.

It can be born in mind that in order to determine the percentage of gold in the alloy, by gravimetric method, it is necessary to affect all the metals present in the sample. A gravimetric method requires to determine the weight of recovered gold from the sample. This can be accomplished by complete dissolution of the sample and then by changing their physical properties from liquid to solid or vice versa, separate them physically by filtration from the liquid, and thus from each other. This particular phase of the analysis is described as recovery, while the purification of the resulting aggregates is sometime refer to as refining.

SEPARATIONS:

The insolubility of gold metal in HCl, H2SO4, and HNO3 affects the separation of gold from the majority of commons metals, only when the metals are in the form of powdered mixture. When alloyed, the separation of the metals becomes a more difficult operation. The basic principle of chemical separations is the different solubility of the materials in their common solvent after their physical state has been changed by a chemical reaction. The physical state can be primarily affected by adding reagents to the solution of the metals, which may combine with gold or with the other elements in the solution and will become a compound which is insoluble in its original solvent. Such an insoluble compound of either one of the elements can be separated from solvent physically by filtration. The completion of such a reaction is very important for quantitative separations. The completion of the reaction is generally affected by heating, or by an excess of the reagent. Separaton of the metals can also be accomplished by oxidation - reduction reactions, where generally formation of other compounds is accompanied by oxidation of metals to their higher state, while reducing the gold to its metallic form, to zero oxidation state. The reactions are chosen in such a way that they can be applied to the strongest solvents of the metal after a substantial dilution, from which solvents the gold can be quantitatively reduced to solid state. Because of the tediousness in making a complete separation from associated metals and the uncertainty in collection of the product the direct precipitation method must be suitably adjusted. This method does not constitute a complete process of purification.

As an example the following method which will be described can produce a proof gold from materials which contain gold in excess of 80% of the weight of the material.

CHEMICAL PREPARATION OF PROOF GOLD

Commercial gold may contain a variety of alloying metals and impurities. The method of making pure gold from any material depends to a great extent upon the character and quantity of materials which may be present in the mass of the material. The method is based on the assumption that the material is of an extreme impurity, however, the total of impurities should not exceed 2%.

The metal is treated in 10 gram portions. The strips of gold are placed in a Pyrex beaker and dissolved with a mixture of 5 ml nitric and 50 ml hydrochloric acid. If silver is present, the hydrochloric acid content may have to be increased during the dissolution of the material. The solution is evaporated almost to dryness and then gently heated until the gold is reduced to metal. Precautions should be taken that none of the reddish auro-chlorides will volatilize from overheating, which may occur as low as 150 C (302 F) and that none of the aurochlorides remain present. The subsequent addition of ammonium hydroxide will result in formation of a yellow precipitate of fulminating gold, if any aurochloride remain in the beaker. This compound is highly explosive even when wet.

The addition of ammonium hydroxide will dissolve most of the silver and copper. After decanting the ammoniacal solution and washing the residue with water, the gold is digested with hot nitric acid. If the solution is wine-colored, the digestion is continued for several hours and re-heated with fresh portions of acid until the absence of color indicates the removal of palladium. (The palladium which is soluble in nitric acid is decanted) and the residue is washed with distilled water. The gold is then re-dissolved with a mixure of 5 ml nitric and 20 ml hydrochloric acids and again evaporated to dryness. The dry residue is now moistened with hydrochloric acid. Then it is diluted with about 800 ml of water. After a faint cloud of the silver chloride settles to the bottom of the beaker -which may occur on standing for several hours or days- the liquid is then filtered through high density filter paper into another Pyrex beaker.

Through the filtrate, a sulfur dioxide gas is passed until the gold is practically all precipitated. The gold is allowed to settle overnight and then the excess liquid is decanted and the residue is washed by distilled water several times. Then residue is treated with aqua regia, and after repeated evaporations down to syrup and addition of hydrochloric acid, the nitric acid is expelled. The product of the second evaporation is acidified by a small quantity of hydrochloric acid and diluted to 800 ml. To this solution, 11 grams of ammonium oxalate crystals are added. The liquid is permitted to remain on a steam bath until the reaction is complete.

The spongy mass of gold is washed now with hot water until free of salts. The gold is then dried and melted in a conical clay crucible (Denver clay crucible) by hydrogen flame torch. Small portions of borax glass are added for fluxing, and after heating the molten gold for several minutes, it may be poured out into a grafite mold. The volume of the ingot now is a half of a milliliter. After the solidification of the metal, the gold ingot is thoroughly washed with hot hydrochloric acid and distilled water. The dry gold prepared in this way may be considered 1000 fine.

It may become obvious to the reader that this tedious operation of preparing the fine gold, lacks practical application for gold recoveries. However, it demonstrates the feasibility of preparation of fine gold by chemical methods. Gold prepared in this manner may serve as a reference standard for the laboratory which uses other more practical methods. Under no circumstances, does it represent a recovery or refining process as used in the industry.

Processes which are applicable for gold in alloyed form with other metals, in excess of 3%, usually are designed in a greatly simplified manner along these lines. Knowledge of the quality of the scrap material and its origin can eliminate some of the lengthy evaporations and long settling procedures. It is important to know, the process can be aimed to one or two separations only, if the associated materials are known, however, one step procedure never constituted a complete purification process or refining. Usually the gold is first concentrated to a batch of impure gold, later purified by one single separation, and ultimately refined by a one - step electrolysis or zone refining.

GOLD CONSUMING INDUSTRY

Consumption of gold in the United States is generally confined to three consuming sectors: jewelry, dentistry and industry. Manufacturers of jewelry consume nearly two thirds of domestic gold. Most of the jewelry is ten or fourteen carat gold. The gold is alloyed with other metals to obtain the desired color and hardness. Jewelry manufacturers recycle 23% of their mint issue as scrap having 77% of the gold going into retail products. This industry operates under a very rigid system of scrap control in order to minimize the loss.

As an example: in 1967, about 14 large jewelry manufacturers used one million troy ounces of gold and returned about 230,000 troy ounces of scrap to the refining industry. In addition they also generated scrap which was remelted in their plants and reused for fabrication of products.

This industry also produces their own alloys by means of controlled additions of major constituents to the base metal, the gold. They purchase the gold in high purity which may be called a "proof" gold, containing more than 99.9% of an Au, a premium product used for standardizing purposes. This purity of gold - so called - " one - thousand fine " is a highly marketable form of gold and is suitable for most purposes, including dental and industrial. The usual grade of refined gold which can be obtained from refiners is 9995 to 9998 fine.

A metal 995 fine is acceptable for international exchange and by the U.S. Mint without a refining penalty. A gold metal in that form usually has the following characteristics:

1. Specific Density at 20°C (68° F) is 19.32g/cc
2. Melting Point at 1063.0° C (1945.4°F
3. Specific Heat at 18° C (64° F) is 0.0312 Cal/gr
4. Electrical Resistivity at 20°C (68° F) is 2.35 micro-ohms/cm
5. Electro-chemical equivalent is variable in chloride solutions.
6. Crystal Structure. Face -centered cubic.
7. Casting temperature. 1100° to 1300° C or 2000 ° F to 2370° F, the oxidizer not required.
8. Mechanical Properties of gold (1000 fine):

Condition	Yield Strength psi	Tensile Strength psi	Elongation in 2 inches %
Cast	....	18,000	30
Wrought	...	......	
Annealed ...	nil	19.000	45
60% reduction .	30,000 ..	32,000	4

As you may notice above, the pure gold may have many qualities, and the means of determining these qualities is out of the range of commonly used methods of identification of gold. A more sophisticated approach has to be taken to determine the fineness of material and its physical constants.

The U.S. Mint set a criterion on acceptance by receiving the gold and paying for the gold by weight resulting from the material which has undergone the Wohlwill process. They received the gold in the form of small ingots or cast bars, weighed it to the nearest 0.1 of a gram, and then placed the gold in their electrolytic process. Before starting, the electrolyte in the process is carefully analyzed, and the weight of the gold determined in comparable accuracy. The cathode is weighed before and after depositing the gold, and then the electrolyte is analyzed again for the gold content. The increase or decrease of gold in the electrolyte is added or subtracted to the total weight of the cathode. The initial weight of the gold deposited on the cathode is devided by the weight of the gold received, and multipied by 1000. The resulting number indicates the fineness of the gold. (This process will be described in Section 2 of this book, under Refining).

This method assures the Mint receiving pure gold and making payments only on gold itself. Any other of the precious metals which may have been present are not considered in price. There is of course, another chemical method which may be used in the preparation of proof gold in smaller quantities.

Apart of gold that originates from the jewelry industry, a great portion of gold may be purchased from the dental industry. The use of gold in industry

Dissolution of carat gold in boiling flasks.

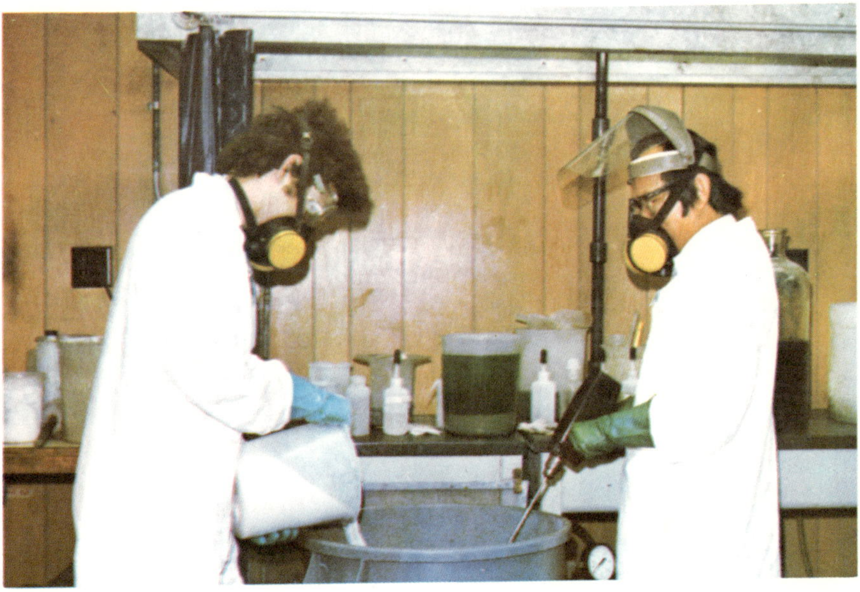

Precipitation of gold

has been practiced for many years. Gold alloys with a high content of gold are substantially different from materials originating from the jewelry business. Gold is used in the form of alloys associated with silver, copper, platinum and palladium. This of course entails some additional processes concerning the platinum group - which consists of metals with close chemical affinities such as palladium, ruthenium, iridium, rhodium and osmium - and therefore require special chemical treatments. The processes are usually hampered by poor dissolution of the material and incomplete separations of elements of the platinum group. Because dental products are more uniform in their gold content, most scrap generated by the dental supplier plants is re-melted and reused in their plants. Only 3% of the total is shipped to the refiners. Still 9% of the total consumption of gold in the United States can be attributed directly to the dental industry. The major suppliers of gold to the dental sector of industry supply about 750,000 Troy ounces per year. In addition, the dental gold is available in smaller batches from dental laboratories and professional dentists, in the form of used gold inlays, partial plates and caps.

A high grade of impure gold is obtainable from small mining operations. This gold may contain a variety of metal elements and therefore its refining may present a greater degree of difficulty. Processes used in mining operations are not comparable to industrial gold recovery processes.

INDUSTRIAL GOLD

The technological advancement in the electronic space and defense industries caused the development of new gold alloys and electroplating processes. The use of gold in electroplating is the largest use of gold within the industrial sector, consuming nearly 80% of all industrial gold. The remaining 20% used in this sector is usually in the form of soldering alloys. All these alloys are specially designed with some specific purpose in mind. Because of this purpose, the ratios of gold to other elements in the solder alloys always varies, not to mention the variance in base metals, soldering temperatures and rate of diffusion. These alloys may specifically be prepared for customers or may have proprietary compositions of elements from the supplier. This of course again, creates difficulties not encountered with metals from other sectors of the industry.

ELECTRO-PLATING INDUSTRY

Gold plated on electric and electronic components is produced by means of electroplating gold from cyanide solutions onto pre-formed or pre-fabricated objects and shapes. The miniturization of electrical and electronic components has become possible because of the unique physical and chemical properties of gold and other precious metals. The consumption of gold in this industry has doubled within the past five to seven years and represents about 28 to 30% of all gold consumed in the United States. The companies consumed more than 1,000,000 Troy ounces of gold in this sector, and it can be assumed that sooner or later all these products - and the gold with them - will be totally scrapped.

CHAPTER TWO

IDENTIFICATION OF THE MATERIALS AND CHEMICAL ANALYSIS

It is understood that the analysis of materials implies that all information concerning the materials origin, appearance, prior processing and specific tests is taken into consideration. Such consideration is necessary to minimize the effort in the analytical work. As an example, knowledge of what sector of industry (as described in the previous chapter) the material originates from, can be the choice among analytical methods.

It also has to be understood that before any prior consideration, the weight of the material (or sample of the material) has to be determined.

SAMPLING OF THE MATERIAL

The sample has to be representative of the total bulk of the material. The purpose of the sample may be to estimate properties of a lot, pile, or shipment, the percentage of some constituents in the lot or the average weight of the shipment, or the specified maximum and minimum content of alloying metals, in this case the gold content of the material. In this case the gold. There are major difficulties in the sampling of a lot or bulk of material:

1. The physical difficulties in separating a representative sample from the shipment;

2. Different problems encountered with differing kinds of materials, which require very specific treatments for each individual case.

NOTE: For statistitical sampling of large lots, a ASTM method of sampling should be adopted.

It becomes evident that an analyzed sample may represent only the material drawn with the sample. In which case, the relationship of the sample to the bulk of the material has to be established by logical extrapolation.

DETECTION OF THE GOLD

As a rule, in treatment of unknown substance, advantage is taken of the solubility of most of the metals and their compounds and the insolubility of gold in one of the acids. In metals or metal alloys that produce colorless solutions in nitric acid gold in the absence of other insoluble materials, produces a black or brownish residue that settles readily, and from which the liquid can be separated by decantation. If the gold is associated with platinum group metals, this residue may become yellowish-brown on heating with concentrated nitric acid. The residue may be dissolved in aqua regia (1 to 5), then diluted, and an additional test can be made to determine the presence of gold. This test is made by adding a solution of stannous chloride containing some stannic chloride.

In strongly concentrated gold and acid solutions, a precipitate of brown metallic gold is obtained. If the solution is but slightly acid and dilute, a reddish purple color is produced by collodial gold and the stannic acid. This color fades on standing. An addition of ammonia produces a red coloration. The sensitivity of this test has been established at one part of gold in 500,000 parts of solution, imparting a perceptible shade.

THE SCRATCH TEST

On jewelry and plated articles, the gold can be detected by scratching the material on the surface of an abrasive stone (touchstone) and then applying a drop of nitric acid to the scratches left on the stone. It is recommended several straight line scratches be made, and that a comparative test be run alongside using a material, which is similar to the tested material and of known gold content. The acid will react with metals on the scratches, and if it is 24 Karat gold on a copper substrate, with silver flush plating, light brownish particles will be left on the scratches. If this test is applied to articles which have been plated, the scratches from first to last, will change in appearance in agreement with penetration of the acid to the substrate material, and the substrate material will remain in the center of the scratch.

If the substrate is copper, there will be a liquid with deepening blue coloration as the nitric acid reacts; if the substrate is steel, the coloration will be greenish; if the substrate is silver, there will be no change in coloration; if the substrate is aluminum, there will also be no change in coloration, but white metal will remain in the center of the lines.

Apart of the exceptions to the rule, if all the lines have the same appearance after the reaction, we can assume the material exhibits uniformity to the depth of the scratches. In such an instance, the article can be tested for its specific gravity.

Determination of Specific Density - Gravity:

Small samples of material, approximately 20 grams in one solid piece, are weighed on the analytical balance to the accuracy of 0.1 of a milligram. A 100 ml beaker containing 50 cc of distilled water is also weighed on the balance. The article is then attached to a fine nylon filament and suspended in the liquid in the beaker. Care should be taken that the article is not touching the walls of the beaker and that it is totally immersed. The weight of the beaker, water, and suspended article is determined. The difference between this total weight and the weight of the beaker with water is equal to the volume of water displaced by the article.

Since we know that one cc of distilled water at room temperature weighs approximately 1 gram, we can derive from the weight of the water displaced by the article, the volume of the article in cubic centimeters: 1cc of distilled water (at 4 °C equals 1gram, and therefore, the density or specific gravity of this water is 1. (1cc/1gram - 1). The weight of the article divided by the volume is the specific density (gravity) of the alloy. For determination of the alloy, we may refer to the table of densities of any gold alloys. An additional tests can performed on plated articles by gently heating a cut off sample from the article in the air. The substrate material which is usually copper will turn into dark black oxides.

GRAVIMETRIC METHODS

I-A. Determination of Percentage of Gold in Jewelry Scrap

This method of analysis is applicable to all materials and gold alloy which are soluble in aqua regia. If any difficulty is encountered in dissolution of a sample,,scch as sample of high concentration of silver, silica or metals of platinum group (dental alloys) then the procedure I-B should be used. In this method the gold is always weighed in pure metallic state and the value of the material is determined beyond a doubt. However, on account of tediousness in making complete separations from associated metals and of uncertainty in collection of the product in a suitable form for accurate weighing, the direct precipitation methods are never used for evaluation of low grade gold bearing materials.

I-A1. Dissolution of the Sample

1.00 gram of finely divided sample (in form of shavings or drillings) is placed into a 250 cc Pyrex glass beaker. 20 cc of 1 HCL 5 HNO_3 mixture of acids is added to the sample, and the sample is gently heated under the fume hood. When the reaction ceases, additional aqua regia is added in small portions, but not to exceed the total volume of 20 - 25 cc of liquid in the beaker. Usually the solutions of jewelry alloys will assume first a yellow coloration which gradually changes into greenish and blue color, with grayish-white silver chloride residue. If the residue remains crusty, then it should be gently crushed and manipulated with a glass stirring rod.

Note* Solution containing platinum metal group will turn brown or if rhodium is present, it will assume a deep cherry red coloration. The treatment of this solution will be described in section I-Bl.

I-A2. After having the solution digested for 20 minutes, the volume will be increased with distilled water to approximately 200 cc, removed from the hot plate, cooled and settled.

The solution is then filtered into 1000 cc beaker and the filter paper (# 40 Watman) is washed with hot distilled water until clear of any discoloration.

I-A3. The filtrate is then diluted to 800 - 900 cc. 20 grams of sodium-meta-bisulphite in crystals are slowly added to the solution while gently stirring the liquid until the crystals are dissolved. Then the solution will assume a bluish tint or if blue already, it will suddenly darken, which indicates the reduction of tri-valent gold to mono-valent aurous chlorides with subsequent spontaneous break down of the solution into brownish precipitate of metallic gold. The shade of the brownish precipitate which may vary from light brown to blackish brown, indicates the purity of the precipitate. If the precipitate appears to be of light beige brown color, then step I-4A can be omitted.

At the first sign of boiling, the beaker is removed to an asbestos pad, let cool and settle overnight.

I-A4. The solution from this precipitation is decanted and discarded. The brown-black residue on the bottom of the beaker is re-dissolved with the addition of 5 cc of aqua regia and subsequently evaporated almost to dryness. To the moist (almost dry) red crystals of auric chlorides, 10 cc of HCl is added, which should re-dissolve all the residue. If the residue is not completely redissolved, then aqua regia must be used again, and the evaporation process repeated. The addition of HCl with subsequent evaporation is repeated two or three times until the crystals appear to be of deep red coloration.

The crystals, after the last evaporation, are re-dissolved by addition of HCl, with a minimum of volume necessary (5 cc). The solution is then diluted to 500 cc with distilled water and the liquid is filtered through #52 Watman filter paper into a 1000 cc beaker.

I-A5. To the filtrate, 5 grams of ferrous sulphate (aqueous solution $FeSO_4$ acidified by HCl) is added and the solution is re-heated to just below the boiling point, cooled to room temperature on standing and filtered through 11 cm #52 Watman ashless filter paper. The residue in the filter paper is thoroughly washed with distilled water, alternately with hot 1% HCl and finaly, washed free of acids.

I-A6. The light brown gold precipitate on the filter paper is placed in a weighed 50 cc porcelain crucible and dried in the oven at 300 °F. The dry sample in the crucible is placed into the furnace and ignited at 950 °F. The furnace should have an inlet for oxygen or atmospheric air to combust all the carbonaceous substances.

68% Gold

The clean white crucible with light beige particles of gold are placed in the desiccator and after cooling to room temperature, the sample is weighed on analytical balances to the accuracy of one tenth of a milligram

Then:

$$\frac{(A - B)}{C} \times 100 = \% \text{ Au}$$

A - The weight of the gold and the crucible.
B - The weight of the crucible.
C - The weight of the samle.
% Au - Per cent of gold in the sample.

I-B. Determination of Gold in Dental Alloys:

I-B1. A very finely divided sample of dental alloy will be dissolved in 1 nitric to 5 hydrochloric aqua regia. However, if dissolution of the sample is not complete, the residue is transferred into stainless steel or silver crucible and fused in 10 grams mixture of 1 part potassium nitrate (KNO_3) and 5 parts of potassium hydroxide (KOH), in the furnace at 450°F to 500 °F. The fusion product is the re-dissolved with water and washed into the original solution of the sample. The solution is filtered through #52 Watman filter paper into a suitable beaker.

I-B2. Separation of Platinum from Gold:

The filtrate from the section I-B1 is treated in the same manner as in section I-A1. However, the solution is diluted to 100 cc with water. Five grams of ammonium chloride (NH_4Cl dissolved in 10 to 15 cc of distilled water) is added to the solution. If platinumis present, then a yellow precipitate will appear after the solution is allowed to stand for minimum of several hours. The yellow precipitate of ammonium-chloro-platinate can be filtered through # 40 Watman paper. The filtrate should be retained for a subsequent determination of gold and palladium. The platinum precipitate is washed with cold solution of 1% of NH_4Cl and 20% of methyl alcohol, then filtered and placed into a weighed porcelain crucible and dried, ignited in the furnace at 1100 °F.

I-B3. The grey spongy platinum metal is weighed calculated and treated as in section

IB-4. The gold is precipitated from the filtrate retained in section **IB-2** which is approximately 125 cc. This solution has to be diluted to a minimum 500 cc and then treated as in section **I-A5** and **I-A6** and calculated as in **I-A7** with the exception of the filtrate from section **I-A5** is retained for the determination of palladium.

I-B5. Determination of Palladium:

To the filtrate from the gold precipitation as outlined under the section **I-B4**, 125 cc of HCl is added and the palladium is precipitated by addition of 10% of potassium iodide solution (KI). Care should be taken not to use this reagent in excess since the precipitate is soluble in its excess.

Heat the solution to nearly boiling, filter and wash free from iron, with 25% of HCl.

Place the precipitate in a crucible, dry and ignite in furnace at 950 ° F, then cool and reduce with a stream of hdrogen or moisten with formic acid.

I-B6. Dry and weigh as metallic palladium applying the same calculations and procedures as outlined in section **I-A7**

Notes* If a sample of dental gold is in solid pieces and cannot be dissolved completely, then the weighed sample can be melted in clay crucible with 5 to 6 grams of copper metal (weighed to the accuracy of sample weighing) and then poured into stainless steel beaker or a dish filled with water. Collect all the dispersed particles in the dish, dry and weigh them, place them into 400 cc glass beaker and proceed as outlined in section **I-B1**.

If the total of dried metal particles doesn't correspond to the total weight of metal put into the clay crucible then correction on the results has to be done in the following manner:

$$\frac{D}{E} = S$$

D - Collected metal particles of sample.
E - Initial weight of sample and the copper in the melt
S - Percentage of the initial sample used in analysis - (divided by 100).

Then:

$$\frac{\% Au}{S} = \text{Corrected results (for loss of sample).}$$

% Au (or any other element) derived from calculations as outlined in section **I-A7** can be corrected for loss in melting.

THE FIRE ASSAY FOR GOLD BULLION

This method is to be used preferably for gold samples with 90 to 100% of gold present.

2-A. Fire assaying is a branch of quantitative chemical analysis in which the metals are determined in metallurgical products (and ores) by extracting and weighing them in the metal state. The metals primarily determined by fire assaying are gold, silver and platinum.

The major operations in fire assaying are usually conducted in box-like muffle furnaces. The separation of precious metals from the lead alloys produced in fire assaying is effected in small vessels known as cupels. The materials of the cupels consists mainly of calcium phosphate, small percentages of magnesium phosphate, calcium and magnesium fluoride and carbonates. These cupels are commercially available in sizes 1.125" diameter, 1.5 " diameter these are the most common. These cupels can be used on delivery if dry. Originally, these cupels or the products for making these cupels were made out of burned animal bones, preferably those of sheep, and then the cupels were made by moistening the 150 mesh of the bone ash product with a small amount of water and then by compressing it into the cupel mold. The mold consists of a ring and a die. The non-uniformity of the cupels may cause an adverse effect on the analysis. This may be the variation in cupel material, the batch moisture, pressure applied when producing them and also the diameter and the depth of the cup. the contour of the cup depression etc.. To generally avoid the non-repeatability of results, the cupels should be kept in desiccators and one batch of the cupels, which is purchased at the time, should be used for analysis and when another batch must be used, check analysis should be made agaist samples of known gold and silver content and thus correct the possible losses due to the cupels. Silver losses in any event will vary from about 1% to 4% for a sample of 300 milligrams. If silver and gold are in the ratio of at least 5 to 1, gold absorbtion by good cupels is usually negligible in most of the cases.

2-A1. Sampling:

Good results in assaying pre-supposes accurate sampling. The sample of gold bullion is in the portion of 400 - 500 milligrams for the assay. The sample then is wrapped with a pure silver foil about 5 times its weight. If silver also needs to be determined, then the silver foil has to be weighed to the significant accuracy of the sample weighing.

In addition this package is wrapped in 3 - 5 grams of lead foil and then cupelled.

2-A2. Cupellation:

The cupellation is a process by which the gold and silver are separated from the lead and other base metals they are alloyed with. In this process the lead is oxidized to litharge which is taken into the pores of the cupel by capillary attraction. This takes place because the litharge is molten at the temperature of the operation.Most of the other metal oxides are infusible at this temperature when in small amounts, however, they dissolved in the liquid litharge and are carried into the cupel.

The cupels should be heated in the muffle for at least 20 minutes before putting in a sample buttons. The cupellations are the best started at 900 °C. The temperature is stabilized at 883 °C. (1621 °F). As a matter of fact the temperature of 800 °C is sufficient for the cupellation process due to the oxidation of the lead to litharge which reaction generates a considerable heat and therefore, the reacting mass of the sample is hotter than the furnace.

The sample button is placed into the pre-heated cupel. As soon as the buttons are put into the cupels the muffle door should be closed. The time of the exposure may vary from 20 to 25 minutes. A good indication of the right cupellation temerature is the formation of solid flakes of litharge upon the inner edge of the cupel, they usually disappear when the furnace is opened. If the sample beads contain platinum metals, it is desirable to raise the temperature slightly above 883 °C through the 21 to 23 minutes of operation. Otherwise, unusually large amount of lead are retained in the buttons.

2-A3. Parting:

Parting is the separation of gold from silver in an alloy containing these metals. This is effected by nitric acid. This acid converts the silver into soluble silver nitrate but is almost without action upon the gold. In order to part readily, the alloy must contain at least twice as much silver as gold. Even at this ratio, it is difficult to dissolve all of the silver. In practice, it is better to have a mch larger portion of silver, except in the assay of gold and silver bullions.

The treatment of the gold silver button from the furnace is as follows: The button is placed in a porcelain crucible or preferably a small 250 cc beaker. Only few milliliters of the acid is needed to start the reaction. This acid should be heated to boiling, and the beads then dropped into it. Solution of the move the last traces of silver. The second acid acid (20% by volume of HNO_3 in water) is always followed by one with a stronger acid in order to remove the last traces of silver. The second acid should be about 1.26 specific gravity or 50% by volume. If a concentrated acid is used first, the gold in the sample button is liable to break up into fine powder which is difficult to manage without loss. By using a rather weak acid first the gold has a tendency to coalesce into a coherent mass which can be washed and weighed as one piece.

After 20 minutes of boiling in the acid, when all the visible action has ceased, the weak acid solution is carefully decanted into a white dish, avoiding the loss of any gold. In the white dish even small particles of gold would be visible. Then about 3 ml of the stronger acid is added and the samle heated to the boiling point. The heating is continued for at least 10 minutes, when most of the silver should be in solution. The acid is then poured off and the gold washed several times with distilled water.

2-A4. Anealing

After a thorough wash with distilled water, the sample is placed in a small weighed porcelain crucible, and dried in the oven at 300° F. Then the crucible is placed into the muffle furnace pre-heated to 1621° F, and anealed for 5 minutes. The anealing causes the brownish-black spongy or fibrous gold to coalesce into a dense flake.

the bead has the characteristics of the yellow color of the metal. The anealing also serves to dry off any volatile impurities which may be present. After cooling, the gold is weighed in a crucible to the nearest one 100th of a milligram, if possible. Then the fineness of the gold is determined as in section I-A7.

2-B. The Assay of Gold and Silver Bullion:

If the fire assay of gold and silver bullion is made, a correction must always be made for the metal losses, because of the great value of the bullion. Standard practice in umpiring dorè bullion calls for running quadruplicates. If isolated samples are being assayed, then two proofs should be run. The samples are usually one gram portions. The sample is wrapped in ten grams of gold (silver - free) sheet of lead and placed into a hot cupel and melted. Ordinarily, one and a half to one and three quarters cupels are used. Up to about 20 parts per thousand the proof will show a loss and therefore necessitating a plus correction.

2-B1. The first parting acid used should contain about 20% of HNO_3 and the second 50 to 56%. The amount of gold after parting is determined and subtracted from the total of the sample. The result then is corrected for the loss of silver.

2-B2. Gold Bullion Assay:

The assay of gold bullion is in principle the same as that of silver bullion. The gold is in excess of silver, however, the process involves the addition of silver and use of stronger acid first than in parting of silver buttons.

I-B3. U. S. MINT METHOD

Sample portions of 500 milligrams are taken for assay. The preliminary cupellation is made as in case of dorè bullion. A ratio of silver to gold, 2.5 to 1 is used in making up the check samples. If no copper is present, then about 3 to 4 per cent of copper is added as it facilitates the removal of the last traces of lead in the cupellation. The cupelled buttons are flattened by hammering, anealed, then rolled, until they are converted to fillets about 2.5 inches long and one half of an inch wide. The fillets are again anealed and rolled up into coronets.

The parting is done by boiling for ten minutes in nitric acid (1.28 specific gravity). The coronets are then washed three times in distilled water, dried anealed and weighed.

The proof usually show a slight gain in weight of gold so the correction is made by subtracting the gain from the average weight of the gold found in the sample portions.

The gold after parting should be in one piece and have smooth edges. The correction must be derived from samples of known quality and same physical form.

3. THE ANALYSIS OF GOLD IN CYANIDE SOLUTION

3-A. This method of analysis is very simple, but very important for control of gold stripping process. The major elements in the solutions are primarily gold, silver and zinc in addition to alkali metal and other insignificant impurities in small concentration. These do not have any effect on final results.

The purpose of this method is to determine how much of the metal is in the batch of stripping solution and if any additions are necessary or just to the contrary; it is important to know if any significant amount of gold has been retained in the solution after gold has been precipitated (or displaced by the zinc out of the solution). Sometimes detection of the gold can prevent great losses in process when the solution is drained and discarded.

This method also may serve to estimate the content of the gold in cyanide-electrolyte when the continuous electrolysis is used for removal of the gold from stripping solutions, and re-circulating the solutions through stripping tanks.

The Procedure:

3-A1. The sample of 50 cc of cyanide stripping solution is poured into 200 cc wide mouth Erlenmeyer flask (graduated in increments of 25 cc, Pyrex brand glass).

3-A2. 50 cc of concentrated sulphuric acid (sp. gr. 184) is slowly added to the sample, UNDER THE FUME HOOD, and then the sample is placed on the hot plate. When the initial foaming ceases, the heat can be increased, and boiling can be continued until the volume (100 cc) is evaporated to 50 cc.

Toward the end of the boiling time the gold metal separates from the acid then coalesces into a spongy button. The sample can be removed on the asbestos pad and cooled to room temperature.

3-A3. The control sample for the completion of precipitation should be examined for solid light brown residue (the acid is usually of light yellow color and very transparent while still warm). If any visible particle is observed, then the precipitation process in the settling tank should be repeated and gold content of the liquid re-checked before draining the spent liquid.

Samples for control of re-cycled electrolyte - stripping solutions will usually contain some gold from previous electrolytical separations and therefore the amount of gold must be determined in order to evaluate the amount of gold in each batch in the process.

3-A4. The acid from the sample is decanted into water while carefully retaining the gold residue in the flask. The walls and the neck of the flask are washed with stream of destilled water until the flask is refilled to approximately 10 cc of water, then the sample is heated to boiling and decanted.

The gold is washed by decantation andboilufseveral times until free of acid and then transferred to weighed porcelain crucible, dried and weighed.

3-A5. The calculations are as follows:

$$\frac{\text{Au grams}}{50} \times 1000 = \text{Au grams/liter}$$

or:

$$\frac{\text{Au grams x 3785.3}}{\text{50 x 31.1}} - \text{Gold Troy Ounces/U.S. Gallons}$$

Either one of the expressions above are important depending, on the volume-calibration of the processing tanks. A good estimate of total volume of the solution retained in processing tanks, pipes, pumps, valves etc., is also very important.

NOTES*: The addition of acid to liquid samples containing high percentage of cyanides is accompanied by violent reaction, gas evolution, foaming and boiling of the liquid. The evolved gas is hydrogen cyanide which is very toxic and deadly. The technicians who are performing these tests should be protected against accidental splashing and splattering of the hot liquid. Preferably they should wear face shields, rubber gloves, and rubber apron. The contact with cyanide liquid should be avoided especially in eyes, mouth and nose. A contact with skin is not dangerous, however, the cyanide should be removed immediately by washing with a stream of water and soap. For precautionary measures and emergency treatment a physician should be consulted prior to attempting to perform these analyses.

A commercially installed fume hood should be used. The exhaust ducts of the installation should be a minimum of 10 feet above the roofs of any ordinary buildings.

3-B. DETERMINATION OF GOLD IN CYANIDE PLATING SOLUTIONS

Although the 3-A method of analysis can be applied to plating solutions in the same manner as described, the large amounts of buffer salts in the plating bath sample renders the break down of the solution by sulphuric acid troublesome and awkward. A more efficient method is described in the following paragraph.

3-B1. To a 50 cc plating solution sample (use 400 cc glass beaker) add approximately 5 grams of sodium cyanide in crystals. The solution of the crystals can be affected by gentle heating and stirring (under the fume hood).

2 - 3 grams of powdered zinc metal (200 - 400 mesh powder shaken into into 25 cc water suspension) is added to the warm solution, and the solution is brought to the temperature of 140°F. The solution is heated while stirring for several minutes and then cooled and settled for several hours or overnight.

3-B2. The liquid is decanted and discarded. The grey residue in the beaker contains the gold and this residue can be treated in the same manner as in 1-A4 through 1-A6 or 3-A2 and calculations determined as in 3-A5.

3-C. DETERMINATION OF GOLD IN CYANIDE SOLUTIONS

VOLUMETRIC METHOD

These methods are applicable to determinations of strength of chloride in gold solutions used in photography, electro-plating, and the electrolyte in the Wohlwill parting process.

The Procedure

In carrying out the determination of gold in cyanide or acid solutons, a measured sample of the solution is freed of oxidizing agents. A measured amount of standard precipitating reagent is added in slight excess of the amount required to decolorize the solution. The sample is gently heated until the gold settles out leaving a clear liquid, a few drops of sulphuric acid may be added and, without filtering, the titration is performed. The gold value of the quantity of the reagent added, minus that, found in excess of reagent, is the gold content of the amount of the sample taken. This method is applicable to the methods mentioned in 1-A, 1-B and 3-B. This titration method eliminates in each of the methods filtrations and the drying time.

3-C1. The sample in form of residue in either one of the previous methods is re-dissolved by aqua regia and converted to chlorides by repeated evaporation and decantation of the liquid and ultimately filtered through #52 Watman paper. The volume of the sample should be approximately 200 cc. Few drops of sulphuric acid is added to the filtrate in 350 cc Erlenmeyer flask.

The filtrate should be warm while titrating. At the first sign of precipitation the titration should be discontinued, and the volume of the titrated solution determined.

3-C2. After complete precipitation and settling, the solution is back-titrated with standard potassium permanganate ($KMnO_4$). Volume of the titration is also determined.

3-C3. Calculate the percentage of gold in the solution as follows:

$$\frac{BC-(A-D)\ E \times 0.065746}{F} \times 100 = \%\ \text{Gold}$$

Where:

A. Milliliters of $KMn0_4$ solution required for back-titration.

B. Milliliters of Ferrous Ammonium Sulphate $(Fe\ (NH_4)_2(SO_4)_2(6H_20)$ solution added to the sample by titration.

C. Milliliters of $KMnO_4$ solution equivalent to 1 ml of the Ferrous Ammonium Sulphate solution.

D. Milliliters of $KMnO_4$ solution required for end point correction.

E. Normality of $KMnO_4$ solution.

F. Grams or milliliters of sample used.

NOTE: $\text{Normality} = \dfrac{\text{Gram-molecular weight/liter}}{\text{VALENCE}}$

3-C4. Preparation of titrating solutions:

Solution A.

Dissolve 3.2 grams of potassium permanganate in 1 liter of distilled water, let stand in dark for two weeks and filter without washing, avoiding contact with rubber or other organic materials. Store in a dark colored glass stoppered bottle.

Solution B.

Dissolve 39.2 grams of Ferrous Ammonium Sulphate dried crystals in 500 cc of cold (5 : 95) H_2SO_4 and dilute to one liter in volumetric flask, with the same acid.

3-C5. Standardiation of the solutions:

Determination of value C. Fill two 50 cc titrometric burets with solution A and B. Run 40.0 cc of solution B into 250 cc beaker and dilute to 100 cc volume. Titrate with solution A ($KMnO_4$) at the rate of 25 - 35 milliliters per minute while stirring slowly. When a pink color appears, let stand still until the color fades (about 45 second). If the pink coloration of the solution persists, then heat the solution to approximately 140^0F. and complete the titration by adding drops of solution A, until the faint pink color persists for 30 seconds. Add the last 0.5 to 1.0 cc drop by drop and allow each drop to be decoloried before the next one is introduced. The measured volume in milliliters of solution A and solution B should be about the same. Then:

$\frac{A}{B}$ = C Value (milliliters of solution A equivalent to 1cc of solution B.)

Example: 40 cc of solution B is introduced into the beaker.

39 cc of solution A is titrated to reach the end-point.

$$\frac{39}{40} = 0.975$$

Then 1 cc of ferrous amonium sulphate is equal to 0.975 cc of 0.1 N of $KMnO_4$.

Determination of value: E :

Determine several times how much of solution is necessary to produce a distinct and persistant pink coloration of the B solution. Usually it will amount to few drops in terms of 1/10 of a milliliter.

Then: Normality of Potassium Permanganate (Solution A).

1 N (One Normal) solution is equal to:

$$\frac{\text{gram-molecular weight } (KMnO_4 \text{ in } 1000 \text{ cc } H_2O)}{\text{Effective Valence}}$$

$$0.1 \text{ N } KMnO_4 = \frac{158.04 \text{ gs} \times 0.1}{5} = 3.16 \text{ grams/liter.}$$

Determination of Normality of the Solution:

ASTM method of standardization: Transfer 0.3000 grams of the National Bureau of Standards sodium oxalate sample, dried at 105 °C., to a 600- ml beaker. Add 250 mls of H_2SO_4 (5:95), previously boiled **for** 10 to 15 minutes and then cooled to 27°C. Stir until the oxalate has dissolved. Add 39 to 40 mls of the $KMnO_4$ solution at the rate of 25 to 35 mls per minute, while stirring slowly. Let stand until the pink color disappears (about 45 seconds). Heat to 55 to 60 °C., and complete the titration by adding th $KMnO_4$ solution until a faint pink color persists for 30 seconds. Add the last 0.5 to 1 ml drop by drop withparticular care to allow each drop to become decolorized before the next is introduced.

Determine the amount of $KMnO_4$ required to impart a faint pink color to the solution by adding the $KMnO_4$ solution to the same volume of the boiled and cooled H_2SO_4 (5:95) at 55 to 60 °C. This correction usually amounts to 0.03 to 0.05 milliliters. A 0.3 grams portion of sodium oxalate requires 44.77 mls of 0.1 N $KMnO_4$.

$$\frac{\text{Weight of sodium oxalate} \times 44.7 \times 0.1}{0.300 \times \text{cc of } KMnO_4 \text{ titrated}} = \text{N } KMnO_4.$$

Determination of Factor Value:

The precipitating value of a reagent and the relative oxidizing value of the permanganate solution can be checked very accurately by adding a measured quantity of the reagent to an excess of gold chloride, filtering, washing thoroughly, incinerating, and weighing the precipitate obtained in a weighed porcelain crucible. We find out that 0.3 grams sodium oxalate will precipitate 0.2943485 grams of gold. Therefore, if we calculate the equivalent of 0.1 N $KMnO_4$ to any of precipitating solutions then we know that every equivalent of 1 cc of the precipitating solution will precipitate 6.57 milligrams of gold. Then by merely multiplying the titrated value of precipitating solution B, by the milligrams of gold, we will obtain the weight of precipitated gold from the sample solution in milligrams.

Then:

44.7 cc of 0.1 N Ferrous Ammonium Sulphate solution when added to the sample will also precipitate the same weight of gold. This volume of the reagent contains 1.7454984 grams of Ferrous Ammonium Sulphate, which weight is 5.82 time larger.

The practical tests indicate that the ratio is 5.96 due to the unstability and hydration of the crystals.

Ferrous Sulphate ($FeSO_4$) has 4.22 ratio etc.

The important standardiation in this method of analysis, is the one, of Potassium permanganate.

5-A

VOLUMETRIC METHODS FOR DETERMINATION OF SILVER

The Thiocyanate Method:

This method is especially adapted to the determination of silver in cold dilute nitric acid solutions. This method provides a rapid determination of silver metal content of silver refining electrolyte. The method also can provide good information on samples with silver content, however, it is insufficient to determine the fineness of the silver bullion.

5-A1. 1 cc of the electrolyte solution is taken from the electrolytic tank by calibrated pipette and transferred to a 200 cc beaker. The volume is made up to 100 cc. Few drops of saturated solution of Ferric sulphate is added as an indicator. The potassium thiocyanate (KCNS) is titrated into the solution until a permanent brownish tint of the solution is obtained. The grams of metal per liter of electrolyte is determined as follows:

5-A2.

$$\text{Grams/liter} = \frac{\text{mls of KSCN titrated x Factor}}{\text{mls of sample}} \text{x1000}$$

The method can be used for determination of solid samples prepared by dissolution of the sample in nitric acid and boiling to expel nitric oxide fumes then diluted to 100 cc in the same beaker and titrated in the same manner.

5-A3. The weight of silver in the sample is determined by following calculations:

$$\text{Grams of silver in sample} = \frac{\text{cc of KSCN x factor}}{\text{Wt. of Sample}}$$

NOTE* Colored salts of cobalt, nickel, copper, if in the sample at high percentage (60%), also Mercury or palladium metals or nitrous acid and chlorine will interfere with the test. The test should be performed only if they are absent.

5-A4. Preparation of thiocyanate reagent:

Dissolve 9.2 grams of KSCN in distilled water and dilute to 1000 cc.

Standardization:

Dissolve 500 milligrams of pure silver (1000 fine) in 5 cc of nitric acid. Heat to boil off nitric oxide fumes, dilute to 100 cc with distilled water, put a few drops of ferric sulphate indicator and titrate with the prepared solution. Then:

$$\frac{\text{Weight of pure silver}}{\text{cc titrated}} = \text{FACTOR}$$

cc (titrated) x Factor = Milligrams of Silver equivalent to 1cc of titration.

5-B

GRAVIMETRIC DETERMINATION OF SILVER BY PRECIPITATION AS SILVER CHLORIDE

For samples which are contaminated with high percentage of interfering element, which could prevent the application of **5-A** method, the following method can be used.

5-B1. 1 gram of the sample is dissolved in 5 milliliters of nitric acid. The solution is diluted to 100 cc and heated to boiling. To this boiling solution 5 milliliters of HCl is added. This is sufficient to precipitate over 1 gram of silver. The precipitate will coagulate on boiling and will remain on the bottom as residue.

5-B2. Filter the residue through Gooch crucible (with a fairly thick pad of asbestos fibre, then wash and dry to constant weight). Wash the chlorides of silver with hot water until free of acid.

5-B3. Finaly, wash once with alcohol and dry at 110 °C to constant weight. Calculate the final weight as follows:

$$\frac{(A - B) \times 0.7625}{C} \times 100 = \%\ Ag$$

A. The weight of the crucible with silver chloride.
B. The weight of the crucible.
C. The weight of the sample.

INSTRUMENTAL ANALYSIS

4-A. A practical application of the instrumental methods is made in the estimation of gold in samples with very small quantities of gold, or in routine work, when the value of gold is expected to be of certain percentage and must be frequently estimated.

ESTIMATION OF GOLD IN WOHLWILL PROCESS

4-A1. A few cc's of electrolyte is placed in a testing tube and a saturated solution of stannous chloride is added. The sample develops a purple color which fades on standing. The test tube is placed immediately into a spectrophotometer and the density of the transmission is recorded at a particular wavelength, with reference to water or a sample of known contaminate. The wave length Lambda is derived from scanning the visible range of the spectrum between 400 to 700 millimicrons. The sample can also be tested without addition of a reagent if in chloride solution and if compared to a standard electrolyte.

4-B1. 0.5 gram of potassium bromate is mixed with 10 to 50 milliliters of cyanide solution. Sulphuric acid is added gradually with constant agitation until the reaction begins. When the reaction ceases, a saturated solution of stannous chloride is added drop-wise until the liquid is just colorless. The tint produced is compared in a spectrophotometer with that from a standard solution treated in the same manner. The later tests should be performed under the fume hood as the toxic fumes may be evolved.

SPECTROCHEMICAL ANALYSIS

4-C. Spectrometry is today becoming accepted as the most efficient method in analytical chemistry in the sense of work performed in a given unit of time. It also possesses a vast capability in determination of multitudes of contaminates in the sample. It also serves as a quick identification of elements present in the unknown sample, however, in the gold industry this method doesn't find too much application due to the fact that it merely tests the surface of the sample. This leaves the analyst with great uncertainty as to the homogeneity of the remaining sample. Since the materials are of great value this method has to be followed by quantitative gravimetric determination of the gold in sample drillings. It is commonly applied to verify the purity of precious metals.

Relationship of measured intensities of spectral lines.

The examination of the sample is done by exposure of the sample to strong electric discharges (sparks). The center of the discharge is in optical alignment with the instrument. The electric discharge produces an excitation field into which the reactants in the electrode (sample) are constantly fed in from the reacting samples at a constant rate, attaining a steady state of dynamic equilibrium. This mutual origin of spectrum excitation and the line intensities is rather related to the luminous column above the sample electrode than to the composition of the sample itself. Then the intensities of the spectrum lines are presented as a function of an apparent concentration of the element in the sample.

The spectrum line intensities are not only a function of the kind and number of atoms of a particular element, but also a function of all other atoms of different elements which were present in the sample (the luminous column of the excitation field).

The analysis derived from spectrometric examination of the sample are only a comparative value against a standard of known values. Moreover, it is easier to determine the low percentages or even small impurities than to determine the percentage of the base metal in the sample. Therefore, spectrometry is not the most efficient method in gold processing industry.

4-D. DETERMINATION OF GOLD ON PLATED ARTICLES

After the initial test which determines the presence of gold on a plated article, it is important to determine the weight ratio of the plated gold and the substrate material. This can be done by following procedure:

4-D1. A representative sample of approximately 5 grams, is weighed on an analytical balance to the accuracy of 1.0 milligram. The sample is attached carefully to the anode, providing for conductivity of each particle of the sample. If several articles in the sample are being immersed in the electrolyte, it can be done gradually by deplating each article separately into the same electrolyte or by combining them together with copper wire. 100 cc of concentrated sulphuric acid is placed into a 250 cc Pyrex glass beaker. The cathode should have a surface of approximately 1 to 1.25 inches square. The cathode can be made out of pure copper metal sheet or stainless steel, or any other high corrosion-resistant material.

4-D2. D.C. current is applied at 7.5 volts. The initial amperage will rise 3 to 4 amps/sq. inch and will remain at this level until the last trace of gold has been removed into solution. A sudden drop-off in the amperage indicates that all the gold has been deplated into the electrolyte.

4-D3. The article and the cathode are carefully removed and washed in another 250 cc beaker with a minimum of distilled water. The washings are repeated until the sample is free of attached particles, and then the washings are placed on a hotplate and boiled down to about 10 cc of volume or less. After reducing the volume of water, which now contains particles of gold, the combined volume is carefully added to the electrolyte in the original beaker. The electrolyte, which is sulphuric acid, appears dark blackish-brown with the gold particles suspended in the liquid. The beaker is covered with a watch glass and placed on the hotplate under the fumehood and boiled for 1 to 2 hours. If sufficiently heated, all the gold will precipitate out of the sulphuric acid and the water which was added will gradually evaporate. The blackish solution will gradually change into a light-brown particulate suspension of gold, and on boiling it will coagulate into one spongy, beige-colored mass of gold. The beaker must be very carefully removed from the hotplate under the fumehood and placed on an asbestos pad and cooled to room temperature on standing.

4-D4. The cold sulphuric acid can easily be decanted leaving the gold on the bottom of the beaker. The sides of the beaker are then washed with approximately 20 cc of distilled water at a time, and then the beaker is placed on the hotplate and boiled for 1 minute. The wash water is then poured off and discarded. The washing of the precipitate is repeated 5 to 7 times in this manner until the liquid is acid-free to the Ph of 5, as determined by Ph paper.

4-D5. The spongy metal precipitate is transferred to a 50 cc Gooch crucible provided with asbestos padding. The weight of the crucible must be determined prior to the the transfer, to the accuracy of 0.1 milligram. The crucible is placed on a vacuum filtration stand and the metal is washed with hot distilled water on filtration. The crucible is placed in the drying oven, adjusted to 150°C (approximately 300 °F) for several hours.

4-D6 After cooling, the dry sample is weighed on the analytical balances and the weight of the crucible subtracted. The net weight of the gold divided by the original weight of the sample and multiplied by 100 will give us the percentage of gold which was plated on the article.

This superior method of analysis is applicable to any plated material on any substrate.

The accuracy depends on good sampling as it has been mentioned, and the minimum of gold in the sample not being less than 0.1%; this in turn is the lower limit for economical removal of gold from gold plated materials.

This method, and any instruments designed for its usage is a subject to a patent claim of the author of this book.

CONVERSION FACTORS

Multiply	By	To Obtain
Amperes/sq.ft.	0.108	Amperes/sq. dm.
Amperes/sq. dm.	9.29	Amperes/sq. ft.
Centimeters	0.394	Inches
Centimeters	0.0328	Feet
Cubic Centimeters	0.0338	Fluid Ounces
Cubic Feet	1728	Cubic Inches
Cubic Feet	7.48	U.S. Gallons
Cubic Inches	0.0043	U.S. Gallons
Cubic Inches	0.554	Fluid Ounces
Feet	30.48	Centimeters
Feet	0.3048	Meters
Gallons U.S.	231	Cubic Inches
Gallons U.S.	3.785	Liters
Gallons U.S.	128	Fluid Ounces
Gallons Imperial	1.2	U.S. Gallons
Grains	0.0023	Ounces (Avoir)
Grains	0.0021	Ounces (Troy)
Grains	0.0417	Pennyweight (Troy)
Grains	1/7000	Pounds (Avoir)
Grains	1/5760	Pounds (Troy)
Grams	0.0353	Ounces (Avoir)
Grams	0.0321	Ounces (Troy)
Grams	0.643	Pennyweights
Grams/Liter	0.134	Ounces/Gallon (Avoir)
Grams/Liter	2.44	Pennyweights/Gal. U.S.
Grams/Liter	0.833	Pounds/100 Gal. U.S.
Grams/Liter	0.1217	Ounces/Gal. U.S.(Troy)
Inches	2.54	Centimeters
Kilograms	35.27	Ounces (Avoir)
Kilograms	32.15	Ounces (Troy)
Kilograms	2.205	Pounds (Avoir)
Kilograms	2.679	Pounds (Troy)
Kilowatts	56.92	B.T.U. min.

Kilowatts	1.341	Horsepower
Liters	0.264	Gallons (U.S.)
Liters	33.81	Fluid Ounces
Liters	2.11	Pints (U.S.)
Liters	1.057	Quarts (U.S.)
Meters	39.37	Inches
Microns	3.9×10^{-5}	Inches
Milliliters	0.001	Inches
Milliliters	1.0	Cubic Centimeters
Ounces (Avoir)	28.35	Grams
Ounces (Avoir)	0.11	Ounces (Troy)
Ounces/Gals.(Av.)	7.5	Grams/Liter
Ounces/Gals.(Av.)	6.25	Pounds/100 Gallons
Ounces (Troy)	20	Pennyweights
Ounces (Fluid)	1/16	Pints
Pennyweights	0.05	Ounces (troy)
Percent by volume	8.0	Pints/100 Gallons
Pints	473	Milliliters
Pounds (Avoir)	453.6	Grams
Pounds (H_2O - 62°F)	0.1198	Gallons U.S.
Sq. Centimeters	0.155	Sq. Inches

ATOMIC WEIGHTS

Name	Symbol	At. No.	International atomic weight 1959	Name	Symbol	At. No.	International atomic weight 1959
Actinium	Ac	89	227	Mercury, hydrargyrum	Hg	80	200.61
Aluminum	Al	13	26.98	Molybdenum	Mo	42	95.95
Americium	Am	95	(243)	Neodymium	Nd	60	144.27
Antimony, stibium	Sb	51	121.76	Neon	Ne	10	20.183
Argon	Ar	18	39.944	Neptunium	Np	93	(237)
Arsenic	As	33	74.92	Nickel	Ni	28	58.71
Astatine	At	85	(210)	Niobium (columbium)	Nb	41	92.91
Barium	Ba	56	137.36	Nitrogen	N	7	14.008
Berkelium	Bk	97	(249)	Nobelium	No	102	(254)
Beryllium	Be	4	9.013	Osmium	Os	76	190.2
Bismuth	Bi	83	208.99	Oxygen	O	8	16.000
Boron	B	5	10.82	Palladium	Pd	46	106.4
Bromine	Br	35	79.916	Phosphorus	P	15	30.975
Cadmium	Cd	48	112.41	Platinum	Pt	78	195.09
Calcium	Ca	20	40.08	Plutonium	Pu	94	(242)
Californium	Cf	98	(251)	Polonium	Po	84	210
Carbon	C	6	12.011	Potassium, kalium	K	19	39.100
Cerium	Ce	58	140.13	Praseodymium	Pr	59	140.92
Cesium	Cs	55	132.91	Promethium	Pm	61	(147)
Chlorine	Cl	17	35.457	Protactinium	Pa	91	231
Chromium	Cr	24	52.01	Radium	Ra	88	226.
Cobalt	Co	27	58.94	Radon	Rn	86	222
Columbium, see *Niobium*				Rhenium	Re	75	186.22
Copper	Cu	29	63.54	Rhodium	Rh	45	102.91
Curium	Cm	96	(247)	Rubidium	Rb	37	85.48
Dysprosium	Dy	66	162.51	Ruthenium	Ru	44	101.1
Einsteinium	E	99	(254)	Samarium	Sm	62	150.35
Erbium	Er	68	167.27	Scandium	Sc	21	44.96
Europium	Eu	63	152.0	Selenium	Se	34	78.96
Fermium	Fm	100	(253)	Silicon	Si	14	28.09
Fluorine	F	9	19.00	Silver, argentum	Ag	47	107.873
Francium	Fr	87	(223)	Sodium, natrium	Na	11	22.991
Gadolinium	Gd	64	157.26	Strontium	Sr	38	87.63
Gallium	Ga	31	69.72	Sulfur	S	16	32.066*
Germanium	Ge	32	72.60	Tantalum	Ta	73	180.95
Gold, aurum	Au	79	197.0	Technetium	Tc	43	(99)
Hafnium	Hf	72	178.50	Tellurium	Te	52	127.61
Helium	He	2	4.003	Terbium	Tb	65	158.93
Holmium	Ho	67	164.94	Thallium	Tl	81	204.39
Hydrogen	H	1	1.0080	Thorium	Th	90	232.
Indium	In	49	114.82	Thulium	Tm	69	168.94
Iodine	I	53	126.91	Tin, stannum	Sn	50	118.70
Iridium	Ir	77	192.2	Titanium	Ti	22	47.90
Iron, ferrum	Fe	26	55.85	Tungsten (wolfram)	W	74	183.86
Krypton	Kr	36	83.80	Uranium	U	92	238.07
Lanthanum	La	57	138.92	Vanadium	V	23	50.95
Lead, plumbum	Pb	82	207.21	Xenon	Xe	54	131.30
Lithium	Li	3	6.940	Ytterbium	Yb	70	173.04
Lutetium	Lu	71	174.99	Yttrium	Y	39	88.91
Magnesium	Mg	12	24.32	Zinc	Zn	30	65.38
Manganese	Mn	25	54.94	Zirconium	Zr	40	91.22
Mendelevium	Mv	101	(256)				

* Because of natural variations in the relative abundances of the isotopes of sulfur the atomic weight of this element has a range of ± 0.003.

[a] The atomic weight varies because of natural variations in the isotopic composition of the element. The observed ranges are boron, ± 0.003; carbon, ± 0.00005; hydrogen, ± 0.00001; oxygen, ± 0.0001; silicon, ± 0.001; sulfur ± 0.003.

[b] The atomic weight is believed to have an experimental uncertainty of the following magnitude: bromine, ± 0.002; chlorine, ± 0.001; chromium, ± 0.001; iron, ± 0.003; silver, ± 0.003. For other elements the last digit is believed to be reliable to ± 0.5.

CHAPTER THREE

GOLD PROCESSING

From the previous chapter on chemical properties of gold and analytical methods, it becomes obvious that the process of collection and extraction of gold from its associated metals will be different for any type of material from which gold is being recovered. The physical form of the material, the composition of the material, the associated metals, and the grade of the scrap will basically determine the extraction and recovery method. In the recovery sector of the gold industry, there are only a few refineries which can recover gold from all types of scrap generated. Only 190 licenses were issued by the office of the Bureau for Domestic Gold and Silver Operations.

Some refineries specialize in recovery from certain types of scrap and semi-process other types of materials for gold recovery at other refineries.

Limiting factors are the technical capabilities to handle and recover gold which is associated with certain metals, gold covered by plastics and other synthetic materials, and gold associated with complex industrial scrap.

Geographical Distribution of Gold
Licenses 1968

New York, New Jersey, Pennsylvania	49.4%
New England	17.1%
Midwest	11.0%
Southeast	5.3%
Subtotal East of Mississippi	82.8%
California	9.1%
Balance of West	8.1%
Total West of Mississippi	17.2%
Grand total	100.0%

Source:Office of Domestic Gold and Silver Operations

The lack of technical capability in the processing of scrap which contains very small percentage of gold in large masses of heterogeneous material is the primary reason for the resale of the material to other refiners. This material, sometimes designated on the market as " breakage" is a giveaway item or " fill " item for scrap dealers to be disposed to larger industrial processors. It is usually associated with other commercial items going to the customer. This type of scrap falls usually below the economical level of some refineries, however, in comparison to ores and other mined materials it could still be considered a rich source of gold. Low grade materials are priced on such a level that the percentage of gross profit exceeds 70% of the price of gold in the material and therefore, they become a rather lucrative business for those industrial complexes which can handle this material. To the contrary, a selected and clean high grade material containing gold in high concentrations, becomes scarcely available and the refiners are bidding on these items the highest possible prices which of course are reducing their gross profits to the minimum. The ultimate price for the gold in the world marketing centers is the ceiling price for the refiner to price his purchase of materials and possible profits. In order to compete, the bidding price usually includes the value of other metals which can be salvaged along with the gold and will place the refiner among the top bidders on the item. This, again, requires a special technology and technical know how that enables him to recover other precious metals or base meterials out of the scrap. Some refiners, not having these capabilities just waive the profit on these associated metals and satisfy themselves with smaller percentage of profit on recovered gold.

This is the reason why the scope of the discussions on the technology in the precious metals industry is very limited. Such discussions should include the operations which are or are not performed, however, still offered to the customer. The refiners also do not want to discuss or disclose the proprietary innovations and methods in industry.

The technical capability or the lack of capability of refiners produced a variety of offers, schemes in pricing of scrap, and even some underhanded activities were experienced by buyers or sellers of scrap. This is due to the fact that a good technology in processing requires a substantial capital investment into their facilities. The processors may experience that the grade of material for which they build new facilities is not available on the scrap market any more and the investment in processing facilities may be written off as a loss. For enterprises which are large enough to tolerate this sort of investment the loss is not too great. In reality, there always will be the recurrence and demand for such a service. To the contrary, this may place the business in better position for competition with others.

For demonstration there could be some examples shown how the industries circumvent difficulties in processing. First, they do not purchase the gold outright, they don't offer directly prices on sampling. They operate on 30 days settlement basis and toll basis. The materials being charged per pound "weighed in" and very small fee for an onnce of refined material. In such a case the refiner wishes to get as much material in, as only possible. If the recovery falls below the expected level, then there is no concern expressed. Sometimes the price of the refined gold doesn't cover the processing cost and the processing cost has to be paid by the seller. In such cases where it is obvious that the material will not bear sufficient amounts of gold, the materials are resold for a lower price to another refiner who may derive profit from the same, being better technically equipped. The pricing and charges for services performed will be dicussed in next chapters.

The american industry as compared to others was affected by an era of gold regulations by the government. Unlike in other parts of the world the possession of gold was not permitted to an American citizen unless he was licensed by the governmental agency with a proof that there is an industrial need for his business. This affected primarily the gold mining industry which produced gold from low grade material at a set price. No efforts were put into the development of new methods because the limitations were in the price of the product and there were no signs of change in marketing for future time. The secondary gold recovery industry also had to work within the ceiling of these prices. However, the price of installations were not so great and advantage could be derived from manufacturing or fabricating of dental and jewelry alloys and products or other usable forms of gold on the market. In reality, each of the refiners was running two independent busness installations due to the fact that the gold products could be manufactured from low priced gold as furnished by the U.S. Mint. Some of the refiners, however, limited themselves to sources of rather "clean" scrap from government surplus or concentrates from small mine operations, and they sold their bullion in rather impure form to U.S. Mint. The fineness of gold accepted by the U.S. Mint fell into a range which could be achieved by rather simple processes. Very few refiners really could produce bullion in marketable form to the private sector of industry, which demands a level of 9998 minimum fineness.

It should be emphasized that each step in processing - as an example - from low grade scrap materials to concentrates of about 60 to 83% of gold and then between these concentrates and 99 fineness level, is a great step in processing, and different technology has to be applied. Between 99 level and 9998 level and over, a completely new capability has to be provided. Such a processing is as much time consuming as the concentrating of the lower grade materials. Even just considering the refinement in analytical work to distinctly and decidedly determine the level of 998 requires a spectro-meter, which price is prohibitive for small establishments. These special techniques are not common in small industrial laboratories. Of course, the spectro-metry technically is just reversing the order of processing by determining the level of impurities in the base material which is the gold.

The scope of this book will establish the requirements for recovery and refining of gold at different levels of operation. There will be an an attempt to describe all the levels of operation if possible, that the reader can select and form his opinion on the type of operation which may be his concern.

It is a common knowledge that mining operations can be done on a individual level by prospectors using a wash pan if any water is available in the desert. On the large scale, however, it requires large and expensive machinery to produce gold out of desert sands. Generally, the rule of thumb is, that the grade of ore or gold bearing materials which are the most commonly to be encountered in processing, establish the level of financial involvement of the operation. It is not how much we want to spend, or save, but it is rather how much money is necessary for successful operation of the mine or secondary gold recovery.

Ideally, a refiner shoul have all the technical capabilities which are necessary. In the practical world, the development of his business to a broader scale will bring about the essential technical experience, and he will acquire gradually the necessary equipment for different processes and will provide himself with suitable personnel and he will ultimately possess all the technical capabilities and thus reaching the level in his business where he will experience much lesser competition within the realm of his activities.

In any case, few nationally wide organizations

which are technically superior can just dominate the market with no serious competition to them and even in some cases they can quote the price of gold which they announce to their competitors that at that price, all the gold they need for their establishment is available. As an example this is generally being done by Englehard Company and Handy and Harmon Co. in New York. Following is the list of principle firms in gold refining industry.

Precious metals semirefiners, refiners, and refiner-fabricators:

American Chemical & Refining Co., Inc., Waterbury, Conn.
Joseph Behr & Sons, Inc., Rockford, Ill.
Cincinnati Gold & Silver Refining Co., Cincinnati, Ohio
Eastern Smelting & Refining Corp., Lynn, Mass.
Edward B. McAlpine, Providence, R.I.
Engelhard Minerals & Chemical Corp., Newark, N.J.
George E. Conley Co., Inc., Cranston, R.I.
Handy & Harman, New York, N.Y.
Lea-Ronal, Inc., New York, N.Y.
MacKay Smelting Co., Inc., Salt Lake City, Utah
Pease & Curren, Inc., Providence, R.I.
Sel-Rex Corp., Nutley, N.J.
Simmons Refining Co., Chicago, Ill.
Spiral Metal Co., South Amboy, N.J.
Stearns Metals Corp., Mount Vernon, N.Y.
United Refining & Smelting Co., Franklin Park, Ill.
Wildberg Brothers Smelting & Refining Co., San Francisco, Calif.

Primary base metal smelters and refiners:

American Metal Climax, Inc., New York, N.Y.
American Smelting & Refining Co., New York, N.Y.

Location of Firms

Some of the principal firms of the secondary gold industry are shown in the table. The major secondary gold refining capacity is in the Northeast, with plants in Connecticut, Massachusetts, New Jersey, New York, and Rhode Island. Representatives of these major firms are also in the Midwest and Far West. Refining capacity has developed in the Northeast because of concentration of consumer industries. Principal karat gold operations are in New Jersey, New York, and Pennsylvania; major gold filled and rolled gold plate operations are in Connecticut, Massachusetts, and Rhode Island; industrial consumers are also in the same areas. Outside the Northeast, major refinery capacity is in Illinois, Ohio, and California. Capacity exists in other parts of the nation, but it is limited, specialized, and localized.

The above table doesn't represent from far the number of companies which are in business of gold refining. Many establishments reaching from medium industrial size of refiners, down to single operation refiners are in the business. As a matter of fact, we can consider everybody who comes in contact with gold to be to some extent in this business.As an example, we could describe a dental laboratory remelting and sometimes chemically recovering their own scrap material or a commercial analytical laboratory which is retaining small 500 milligram samples from their customers, which were the samples used in the analytical work. They may ultimately collect a batch of gold. Another example can be described. A dental technician who may collect a few ounces of gold from the patients extracted teeth and bridges, which he may sell or if "clever" enough, refine in his laboratory.Another example, which frequently can be observed. A metal scrap dealer melts his gold scrap into an unsightly blob of impure metal, claiming for himself an unreal value which he never established by any testing.

Another poor example of gold handling is seen

in the case, when electronic scrap which was submitted for analysis was offered at a purchase price so low, that any scrap dealer having seen the articles brilliantly shining, offered a better price and re-sold the lot to another profit seeking person. With small profit always added in each and several steps of re-sale, the material gained full price of good gold bearing scrap. The moment of truth came, however, when the last re-seller contracted a refiner and had to endure a tremendous loss in his investment. As a consequence, the whole of the precious metals industry suffered a bad reputation at the mouth of this man.

This aforesaid article is leading to an important conclusion: The real value in this business is not in the speculative price of the gold, but in the technical knowledge and the technical capability to perform the recovery of gold in multi-level processes. The capability to segregate and separate a small batch of gold and settle a fair price with the owner and at the same time have several other processes running which may be even on continuous contracts. It is the capability to pay the top prices with the minimum of losses to the customer and thus gain the confidence of the customer, and through this effort to secure the continuous supply of gold bearing materials.

THE BASIC CONSIDERATIONS

It can be generally assumed that a prospective enterpreneur in precious metals refining formed his idea about the general market of gold, and possesses some experience in the metal industry. It may be either the manufacturing or just metal marketing that he feels he may find himself a place in this competitive business, and that it would be to his advantage to provide these services to industry and that there is a demand for his specilized technical services in his area. Or, it may be that he is already engaged in fabricating of jewelry or other salable articles manufactured from precious metals. The refining of gold and other precious metals is just a desired technical capability which he doesn't have in his present business and through this addition of technical knowledge to his business he could further his aims in his line of business.

Generally, the scrap that is available in abundance is in the form of gold plated articles or electronic components, therefore, an important process for any sort of enterprise will be the extraction of gold from industrial scrap originating from plating industry. In subsequent articles we will describe the basic process in gold recovery from electronic scrap and the requirements for establishing a plant for this particular type of processing.

THE PLANT

Location:

Under any circumstances, the analytical laboratory and processing plant must be located in the industrial section of the city, usually under the city code M-2.

Building:

The building facilities should have an access from the street for cars and trucks, an area for loading and unloading of shipments. Preferably a car entrance into the building and parking facilities. The building itself should have cement floors uniformly converging to drainage in each laboratory or processing facilities.

Utilities:

There should be an adequate cold and hot water supply in each of the sections of the building. The electric industrial current supply should be provided for at least 440 volts and 200 amperes, transformation into 220 volts and a subdivision for 110 volts throughout the outlets in the building. The power supply lines in the facilities, if rented or owned, should be put into non-corrosive conduits.

Security:

The entire facility should be capable to be secured with locks and an interlocking system for safety. A small floor vault should be provided in laboratory space or office for temporary holdings of precious metals. The grill work in the windows should be hinged and locked in such a way that it protects the plant, but doesn't prevent the people inside of the plant to leave the building in case of emergencies. There should be at least two emergency exits from the plant. The plant should be also provided with reasonable fire prevention facilities, such as fire hoses and extinguishers.

Ventilation:

A very important aspect of the building is that all the facilities are located on one single level floor, a ground floor, which has the inherent quality to allow for building a multitude of exhaust and ventilation systems with proper air-intakes. The air exhausted from the building has to be replaced by a fresh air-intake without the danger of re-circulating the spent air.

Handling and Storage Area:

The handling area should be located adjacent to the parking and unloading space. No scrap storage should be allowed in the processing or access areas of the building. An adequate storage room should be planned for the operations.

Office Space:

Office and the laboratory should be provided with air-conditioning. The air intake must be independent of the facilities

Storage of Liquids:

Outside storage tanks should be provided for spent cyanide solutions and a separate tank for an acid waste solutions. These tanks should be located outside of the buildings where they are accessible to the trucks and also to the personnel operating the facilities, however, this area must be out of the reach and the access of the public.

An inside storage tank for minimum of 100 gallons of distilled water must be also provided with an access for refill hoses from a truck which may be parked within 50 feet distance from the tank. This storage tank is important even in case, when the distilled water is produced in the plant.

PROSPECTING FOR THE LOCATION

Each of these conditions should be kept in mind when prospecting for the business location. These conditions will be discussed in detail as to the fitness of the facility for each of the particular processing involved. It is not necessary to mention that this expected from a medium sized small business plant. However, some of the processes can be carried out as small batch refining in the laboratory, not necessitating any industrial facilities. In this book, it is assumed that the best description can be presented on the scale of at least $750,000.00 yearly gross. The reader has to be aware that the smallest process of gold recovery is in reality the process of chemical gravimentric analysis described in the previous section of the book. Magnification of the processes and the projection of the processes to this medium sized industrial business, entails the introduction of industrial techniques and process equipment design on every level of magnitude. The same conditions however, for chemical reactions can be carried or simulated in the laboratory. As a matter of fact the best analysis are those which are chemically closest to the industrial processing and reflect directly the recoveries which can be obtained even when the over-all accuracy may suffer.

the plant
access from the street

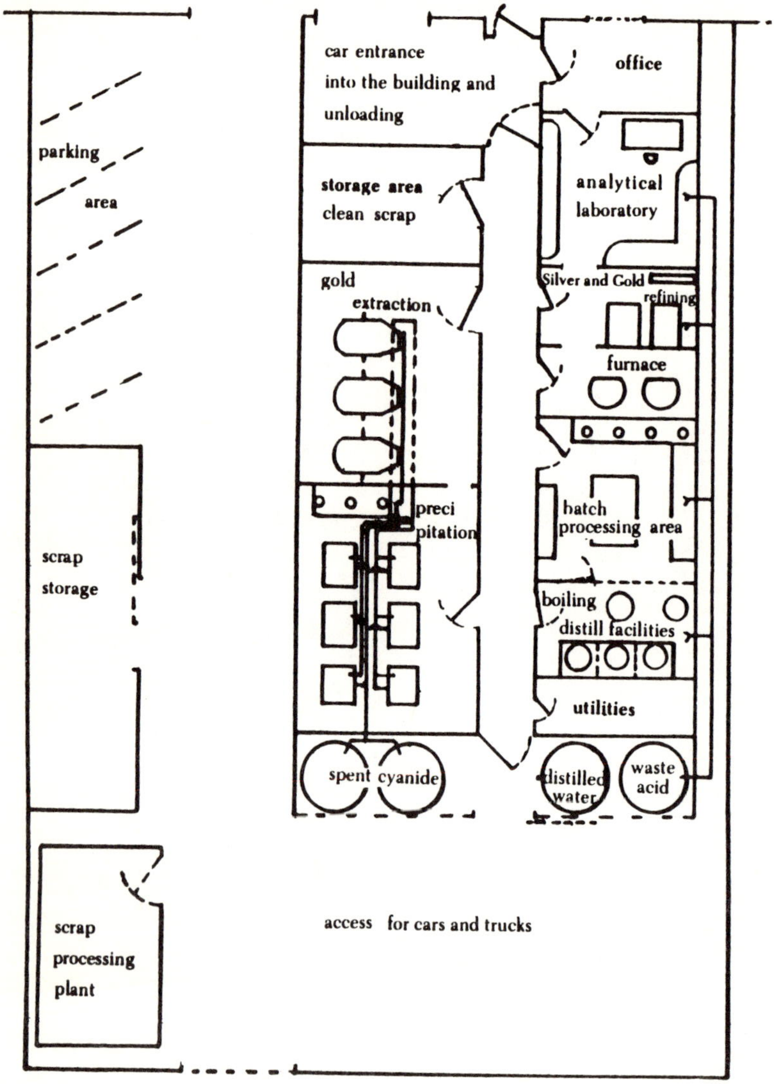

FACILITIES

Two facilities (shops or assigned space within the building) have to be provided for the cyanide process. This working space has to be physically separated from the other facilities in the building because the handling of the cyanide could cause undesirable contaminations of the other working spaces.

THE CYANIDE STRIPPING ROOM FACILITIES

The stripping room facilities should not have any furniture or permanent installations. The tumblers which may range from small ones to large ones, are aligned in the center of the room, parallel to the walls, and accessible from both sides. The most suitable tumblers for this operation are the steel tumblers of the cement mixer type. The ventilation of the room is provided by a single fume hood running overhead and above the mouths of the tumblers. The air exhaust should be at the rate of 250 cu. ft. of air per minute at each tumbler.

The tumblers should have sufficient freedom to be swung over to the other side of the access. This side is not provided with an overhead fume hood and thus allowing full and unrestricted access for loading and unloading the scrap. The loading of the scrap is done by tow trays which are brought in from storage on a push cart. No temporary or permanent storage of scrap should be allowed in the processing room or adjacent corridors. The removal of the scrap from the tumblers after processing is done by lowering the mouth of the tumblers and rotating them in reverse and thus dumping the scrap directly on to low profile trays provided with casters and drainage hoses. At the removal of the scrap from the tumblers there is always liquid remaining on the clean scrap and should be drained on the trays. Several of these carts should be available. The processed scrap should be discarded outside the building.

The processing site of the tumblers which is provided with the overhead exhaust system should also be completely free of furniture and other installations, except for the control panel which is attached to the hood.

The processing of the scrap is done with open tumblers for these reasons:

1. During the processing there is a constant feed of hydrogen peroxide.

2. It is necessary to determine visually, the completion of the process.

It is desirable that adequate lighting is provided above each of the tumblers.

During the operation, the tumblers will exhale vapors and liquid mist which is drawn into the exhaust system. A transparent shield hanging from the hood will prevent the liquid from accidentally splashing into the room. The shields are suspended in appropriate height above a trough running along the tumblers, placed in such a position that the liquid can be poured into the troughs from the tumblers. The trough itself is made from stainless steel or PVC or high-density poly-propylene. The liquid is poured into the trough by slowly lowering the mouth of the tumbler until all the liquid can run out. For this purpose it is desirable to have a motor driven tumbler, as it is difficult to control the slow tilting of the vessel, which is laoded with a large volume of liquid. To prevent the scrap from falling into the trough, a screen is attached to the mouth of the tumbler.

The liquid loaded with gold is collected in the next room in a filtration tank. This means that the tank is provided with upright filter columns or plates and any exit from the tank to the pump is passing through this filter. The filter is rather coarse, however, not less than 100-300 microns. This collecting and filtration tank is used as a sump tank for the pumping system. The outlet from this tank and the pump has to be in such a position that most of the liquid will be drained into the tanks after pumping. The content of the collecting tank doesn't have to exceed the content of either one of the tumblers. The tumblers have to be placed in such a height that all of the liquid from the troughs will run into the collecting tank. The outlet of the filtering system is provided with two gate valves, one which feeds the liquid into the pumping system and the other one for draining the wash liquid.

On the processing side of the tank, there is only one cart available for the processor to bring in preweighed batches of cyanide and calcium oxide.

For safety of the personnel, there should be provided an emergency shower or facility for washing. The processors should be equipped with protective shields, rubber aprons and rubber gloves. This personnel safety equipment should not be removed from the room at any time. Suitable hangers should also be provided in the facilities. There should be provisions for emergency treatment and an exygen supply for each operator. No smoking should be allowed in the facility. Bad smoking habits may be a transportation vehicle for the cyanide to reach a persons mouth. Anything that is accidentally dropped on the floor should be unconditionally discarded or washed. The suspended shields from the overhead fume hood should always be in position to prevent splashing of the liquid. They should be removed only when an intervention in the process is necessary.

An automatic dispenser for hydrogen peroxide is mounted above the rotating mouth of each tumbler. This liquid also is very dangerous for burns which may result from contact with the skin. Refilling of the dispensers has to be done with utmost caution, the mounting of the dispensers must be absolutely safe.

THE PLUMBING

The plumbing for supply of hot and cold water is mounted above the equipment on either side of the access. Short hoses are suspended from the plumbing which enables the operator to fill the tumblers or spray-wash the scrap in the tumbler. Each room is provided with a floor drainage.

FEATURES OF THE STRIPPING ROOM INSTALLATION

It can be noticed from the description of the installations and the drawings that all equipment is moveable and therefore, the floors can be kept clean of liquids and contaminants. The operator doesn't come into contact with the liquid during the operation at any time except when some adjustment is deemed necessary. The loading and unloading of the tumblers can be done anytime during the processing. The collection and direction of the liquid to the precipitating tanks does not require an additional help in processing. Automatic features such as hot water metering and hydrogen peroxide dispensing, and motor driven tilting of the tumblers enables the personnel to work at high efficiency and therefore, it lowers the cost of processing. The safety in handling cyanide is very good.

PRECIPITATION TANKS FACILITY

The room adjacent to the processing facility, is equipped with precipitation tanks. The volume of the tanks may vary for each tank if individual batches of scrap are intended. It is necessary to have at least six 75 gallon tanks and also some larger ones, such as 200 to 300 gallon tanks. When a process control is intended, than this room may have a series of tanks of equal volume, six 200 gallon tanks. These tanks are standing in an upright position, in the center of the room parallel to the walls. This position allows the operators access to each of the tanks for drainage.

The plumbing for the transfer of the processing solution runs through the center between the tanks, with hot and cold water outlets for each tank. Above the plumbing installation, there is a platform on which the processor can reach the top of the tank to make necessary additions.

These tanks are covered completely with two fume hoods running parallel to each other, with an opening in the hood for the addition of chemicals. The tanks are provided with a liquid level indicator, a thermometer, and slowly rotating stirring paddles which operate independently.

The liquid is heated or maintained at a temperature of 140 °F. by electric immersion heaters jacketed by quartz. The process liquid is fed by pumping from the collection tank over the top of the tanks into the precipitation tank. Junction of all the plumbing for each precipitation tank should be provided with a valving system that each tank can be filled at one time, and then the liquid diverted to another tank. The immersion heaters must reach almost to the bottom of the tank and the heating elements should not reach higher than half the volume of the tank so that the tank may be operated without being completely filled.

The tanks are usually made of steel and can be lined with poly-propylene or PVC. The draining of the tank is provided through the side of the tank into a manifold and another re-pumping system is used to transfer the liquid to the outside storage tank. The intake of this drainage system is provided with the pipe inside to reach the bottom of the tank. The drain pipe is fitted out with a screen with an open top that most of the liquid can run through the top until the liquid reaches the screen and then the screen prevents the precipitate from being pulled into the drainage. The chemical disposal tank located outside should be rated at 3000 gallons each.

The bottom of the precipitation tank is provided with an opening for removal of the residue. The opening is large enough that the operator can use hand tools to flush the residue into a tray. Also, it has to be taken into consideration that the bottom of the tanks are provided with slanted sides converging the liquid to a drainage hole in the side. An installed shower in the upper part of the tank may provide the necessary shower to wash down the walls of the tank, but this can also be achieved by having a hose available at the platform between the tanks.

The trays into which the residue is discharged should be large enough to hold some liquid. These trays will have to be lifted on a table in order to drain the residue into glass flasks.

BOILING FLASKS

The round bottom boiling flasks made out of pyrex brand glass are supported by a ring in a heater mantle. These boiling flasks are then mounted on casters. The minimum volume of flasks is 50 liters, but not more than 100 liters for handling ease. It is very important that each of the boiling flasks neck is mounted approximately at the same height from the floor and has the same width of neck. These flasks have to be fitted under the funnel type dispensers and also under other exhaust systems in different facility.

THE GOLD EXTRACTION PROCESS

THE CYANIDE STRIPPING

For most of the materials containing gold, this process is very important and is described as a major chemical process for concentration and gold recovery from industrial scrap.

In this process, the material is tumbled in a sodium cyanide solution and then allowed to precipitate by inc dust. The zinc is dissolved from the residue by HCl and is discarded. The residue is treated with aqua regia and gold is precipitated from its solution by sodium bisulphite. The residue from this precipitation can be cleaned by washing, then melted into electrodes which are at that time about 99 fine. These electrodes are used in Wolhwill process for refining. Incidental concentrations of silver which are in the form of silver chlorides can be also recovered in subsequent process.

THE CHARACTERISTICS OF THE PROCESS

As it can be discerned, this process is used for stripping of gold from plated surfaces of any kind and from any kind of substrate material. It is a process of concentration of gold from vast colume of materials originating from industrail scrap, mining operations and etc. This process is not designed to dissolve solid pieces of gold or thick layers of gold filled material.

Most of the scrap doesn't contain materials in solid pieces, they usually are separated in different batches by the industry from which the scrap originates. It should be understood that "most of the scrap" is just "scrap" and not gold. Gold on this type of material is finely divided over a large area and voluminous scrap materials of all sorts and has to be concentrated into a workable volume for processing.

1. RECEIVING THE MATERIAL

The scrap material usually comes in steel barrels or cardboard boxes. The material has to be received "as is" and weighed. The receiving room is provided with a scale in the range of 1 to 200 pounds and an additional scale of 1 to 1000 pounds.

A barrel full of liquid weighs approximately 400 to 450 pounds. A heavier weight in the barrel is hardly to be expected, because it would pass the capability of normal industrial help to handle compact packages. Unless larger containers are expected to be received, the receiving platform can be equipped just with a push cart.

2. SORTING THE MATERIALS

Subsequently to receiving the materials, material has to be dumped out of packaging, and roughly assorted into tow trays. The original packing has to be re-weighed and the net weight of the material determined.

A receiving slip has to be issued with a copy going to the office. The tow trays should be pre-weighed and marked in visible places. This material is then placed into a secure storage room and will be processed only on directions from the foreman.

When the material is received in the plant storage room, it must be thoroughly examined. On many occasions the material contains not only gold materials but it may contain also items which may even interfere physically with the processing. Example of such material is: paper bags, paper wraps of any kind, and plastic bags. On occasions, the electronic material is still in its original packaging which is usually waterproof and could prevent the chemcials from reacting on the gold plated surfaces. Attention also has to be paid for materials which have been lacquer-sprayed or epoxyed or varnished. Such materials have to be taken out of the batch for special treatment. This usually happens because many people in other industries believe that gold processing is done by firing the materials. This is not true.

Also, it can be found that other types of material have been added by mistake, such as steel bars, copper pieces, zinc casting alloys, etc. Zinc casting alloys (white metal) may completely prevent the chemical reaction in gold stripping tanks.

Hand sorting of the materials, under the supervision of the foreman, is very important for determination of what process shall be used for each of the assorted materials. For instance, electronic components which may be completely enclosed in their plastic or bakelite mountings, such as plugs, connectors and others, have to be processed in separate batches, and have to be physically crashed, tumbled in ball mills, shredded or the plastic material has to be removed by combustion.

The assorted material in tow trays have to be provided with processing slips for control. The processing slip for each batch must carry the account number of its origin, the net weight, the date of the shipping, and the number of the packages in which it was received. These prepared materials for processing will be used only when scheduled for processing by the foreman.

The allotted material for different processes is taken to the processing rooms. The plated material which does not have any plastic parts, like pins, clips, etc. as compared to gold plated wires or electronic circuitry boards may be loaded separately into the tumblers. The metallic material can be loaded in much larger weight. The only important criteria is that all the material, when tumbled has to be immersed in the liquid.

Solid pieces of gold such as shavings, or drillings must not be used in stripping process because a complete dissolution of material cannot be expected at the same time with complete removal of thin layers of plating. If attempting to process dissimilar materials in the same batch, it will result in incomplete dissolution of gold in solid pieces and heavy plates, while over exposing the other materials to the stripping solution. This will cause a severe contamination of the solution. The metals such as copper, or brasses (brass contains up to 35% of its weight in zinc metal) will displace the gold out of the solution in form of a precipitate which will never reach the precipitation tank, being deposited on the filters of the transfer system. In this manner, a substantial loss of gold in the process can be experienced. Moreover, later segregation of the material, which was not completely stripped in the process or dissolved, is more difficult after the tumbling, than it was to do the work right in the first place.

3. PROCESSING OF THE ALLOTED MATERIAL

The material is brought to the processing facility in the tow trays on push-carts. The material is loaded into the tumblers from the side where no overhead exhaust fume hood is preventing the access to the equipment.

The weight of the material, which may be combined from several trays, is marked on the processing slip. It does not matter how many pounds were dumped into the tumbler, however, it is important to register the following:

1. The origin and account number.

2. The total weight in the tumbler. The distribution of the lot and the weight of each kind in the shipment.

3. Date of processing.

4. Volume of water in the tumbler.

5. The tumbler is then filled with hot water from the overhead faucet. The tap water should be 140-150°F. temperature. The tumbler should be filled to a calibrated line. The volume is determined by metering the supply line.

6. The tumbler is then swung over to the processing side. A power mechanism on the tumbler is desirable for this kind of operation.

7. To the liquid, an ounce of Calcium Oxide(CaO) powder is added, and the tumbler allowed to rotate for several revolutions.

8. To the alkaline solution, 8 o. of $NaCN$ crystals

per gal. of water are added from pre-weighed supply containers. The crystals are readily dissolved after the initial revolutions of the drum.

9. Initially, about 200-300 cc of hydrogen peroxide is added to the solution from the overhead dispenser. The rate is adjusted to approximately 1 liter per hour feed in. The reaction of sodium cyanide with metals other than gold may have started by the addition of the chemical to the liquid. However, the dissolution of gold metal starts by addition of the hydrogen peroxide while other metals become rather inert in the oxidied solution.

10. This process is allowed to continue until the underlying material will be stripped of gold. The solution in the oxidized state becomes gold-yellow to brown, but transparent. If the material is not completely stripped in 10 to 20 minutes, then the solution is dumped into the trough and the treatment repeated.

11. Before decanting the solution, a sample of 100 cc or more is taken and the processing slip from the tumbler is attached to the samples. The samples are taken to the laboratory for analysis which determines the contents of each batch. Hence, the importance of the weight of material in each processed batch of scrap and the metered volume of water in the processing equipment.

12. The control analysis in the first stage of the process are done by 3A method described in previous chapter. The results are accounted for each batch and origin of the material for the customer and also for the total recoveries of the process. Periodically, a break for accounting purposes is determined by the foreman or accountant. If the total in recoveries do

not coincide with the sampling analyses then accommodations have to be made for process and personnel efficiencies. The averages of wight of gold loss in process have to be determined and accounted for.

13. The solution from the tumbler is slowly poured into the trough which is located along the equipment. There should be only one of the tumblers decanted at a time and the collecting tank should not over run its capacity. Prior to any manipulation of the liquids, a routing for the solution should be determined. The tumblers should be provided with a temporary screen which can be attached to the mouth of the vessel to prevent accidental spillage of the scrap into the trough, also the trough may have a screen attached at the place where the solution is decanted.

14. The remaining scrap in the tumbler is showered by cold water and rotated several revolutions. The first two washes of small volume are added to the solution in the collecting tank, then the solution is pumped-filtered into the precipitation tanks in the next section of the facility.

15. The scrap in the tumbler is repeatedly washed several times with cold water and the wash dumped in the trough and diverted to drain. The tumbler is then swung over for unloading of the stripped but clean scrap material and subsequently re-loaded with additional material, following the same steps in procedure.

16. The solution (still hot) is transferred into the precipitation tanks. The tanks are filled, one at a time, and heated to maintain the level of the original temperature of 140°F. No control over the volume is

needed at this time.

17. To the cyanide solution in the precipitation tank, sodium cyanide is added. Approximately 2 to 3 ounces per pound of zinc is added.

18. After allowing a short time to dissolve the crystals of the added sodium cyanide, a suspension of zinc dust (shaken into the water), is added in portions of 1 to 2 pounds at the time. After initial reaction and foaming ceases, the solution is allowed to precipitate. At this operation, the solution is constantly stirred by motor driven paddles.

19. A second addition of sodium cyanide is needed if the solution is still reacting or the coloration of the precipitate is brown. The small (1 pound) additions may be necessary until the greyish particles in the solution are observed. The solution is maintained under same condition for half an hour. Then the heating and stirring action is switched off and the solution is allowed to cool on standing and settle over a 24 hour period. The foreman determines the time of draining as the minimum time for settling is 10 hours. The working schedule is rather of more importance. An excessive standing of the solution may cause loss of gold by re-dissolution in the liquid. However, a deposit of zinc powder on the top of the gold precipitate prevents such losses.

20. After settling the tank, the solution is drained into storage tank outside the building and the walls of the tank are washed by shower of cold water to remove particles which may adhere to the wall of the tank. The excess of the liquid is drained into the waste storage tank. The wash is repeated once or twice.

21. The residue in the tank is removed by first completely draining the bottom of the tank, and then opening the large cap at the bottom of the vessel. The grey-brown residue is transferred into a low-profile tank which is mounted on weels or casters and transferred to the gold processing facility.

The above described processing of gold bearing scrap is considered for concentrating process where exposed surfaces of the material are stripped of gold deposits and segregated from the bulk of the scrap material. The reaction in this process does not generate toxic fumes or gasses. However, the agitation of hot solution and the addition of hydrogen peroxide causes fast evaporation of the liquid (water evaporation) and also the hydrogen peroxide rapidly decomposes in the hot liquid with evolution of oxygen. The gaseous phase is not toxic. The minute particles of the liquid in the atmosphere above the solution are carried out from the surfaces in form of mist. The sodium cyanide liquid is highly toxic and may reach a deadly concentration in the air above the processing equipment. The fume hood must have provision for condensation of these vapors, such as scrubbers or condensation louvers.

AUXILIARY OPERATIONS AND FACILITIES

The cyanide stripping process as described is a universal method for concentrating gold from large quantities of various materials. The most important application is, of course, in handling of industrial scrap materials from electronic and plating industry. However, this process is applicable to almost any kind of operations, if the facilities and handling procedures are adjusted to a specific project. This chemical process is capable of extracting gold from sands, ores, weldments, low-grade tailings from previous operations, ashes of combusted materials, etc. We may find it undesirable to contaminate the facility with these materials and lower the efficiency of the established operation. In the course of business operations, there will be frequent demand for special services and unexpected propositions.

In the realm of projecting the installations, a section of the building or part of the open-air yard space should be assigned to special projects. The nature of these projects will be described in the following paragraphs.

PHYSICAL AND MECHANICAL HANDLING OF THE MATERIAL

Some scrap material originating from electronic and plating industry are assembled and locked in their mountings. It is determined by chemical analysis that the material covered by plastic or bakelite contains a portion of the total percentage of gold estimated in the lot.

It is very important from the point of economy to determine what portion of the total can be recovered by special operations. Considering the cost of the operation and the value of the material after recovery, it is sometimes not fully justified. Usually when the contents fall below one-third of the total recovery, such operations should not be considered practical. Special thought should be given to the fact that the corrosive action of the cyanide solution reaches some times into some depth along the protruded metal particles and the total of the recovery can be achieved by simple washing action omitting additional treatment.

The material can be crashed, shredded or pulverized in a ball mill, then washed by flotation. Sometimes a roasting of the material is necessary prior to these operations.

A special outside facility can be reserved for these activities. Under no circumstances should such activities be carried on in the building of the refinery. More damage can be experienced on the equipment than the price of partial recovery of the gold can warrant.

Combustion of the materials can be carried out in the ovens. First at 590°F., then increasing the temperature to approximately 700°F., and 950°F. The escaping fumes are re-cycled through after-burner and combusted at 1350°F.

Combusted materials are shredded to powder and then roasted at 950°F. with access to oxygen. Remaining of the oxidized materials can be smelted with additions of copper. These melted metals are poured into water through a dispersing screen and finally dissolved in acids. This material can also be treated in the cyanide process depending on their physical condition. If the material originates from plated articles then it may be successful. Concentration of the metallic particles is necessary by flotation and washing prior to cyanide processing.

EXAMPLES OF LOW GRADE ORES AND STRIPPING WASTES

Cyanide recovery of gold from low grade scrap materials or small mining operations presents a problem of handling large quantities of material with a minimum of effort or expense.

For purposes of handling of some materials with insufficient concentration of precious metals, a following procedure can be adopted.

Material is unloaded into an auxiliary tank which does not have to be provided with fume hood. A very low concentration of Sodium cyanide (0.1% or higher) is circulated in a tank of large volume of liquid. The solution is kept alkaline by addition of calcium oxide. The addition of the cyanide can be done by first estimating how much is needed for the total recovery. Then the cyanide crystals are dissolved in a small volume of water and boiled with Sodium peroxide. This concentrated liquid is then added as required to the large volume of the circulating solution.

The recovery of the gold and other heavy metals is affected by circulating the solution through charcoal filters. The charcoal filters can be washed with hot alkalies, such as sodium hydroxide and the gold reduced by zinc from the alkaline solution. This process does not require any attendance, however, a check must be performed systematically in time intervals of one day or longer.

A similar process has been investigated by the Metallurgical Research Division of Bureau of Mines.

Gold in this manner can be recovered from low grade ores and stripping wastes.

SOLDERED MATERIALS

Aerospace and other industries use substantial quantities of gold and silver soldering materials for the manufacturing of special devices. These materials are sometimes offered for bidding. Preliminary examination of such material AND COMPLETE EVALUATION IS VERY IMPORTANT.

The soldering materials are usually applied to the structure in very uncontrolled manner. The total of materials claimed by the manufacturer is established by weighing the structural materials before and after the soldering operation. This, of course, is not sufficient information for committing substantial investment for the purchase.

At some occasions, solders of different compositions are applied in several steps of the manufacturing process. For instance, a combination of gold 80% - nickel 20% solder and in subsequent process a silver 80% - Palladium 20% may be applied.

To resolve this combination of elements fused into stainless steel structure is a special project for recovery of precious metals. Special handling and process facilities must be adapted.

Application of the cyanide extractive process to the structures welded in this manner may not always be successful due to several reasons. The soldered material is diffused to some degree in the structural material and only a prolonged leaching by stripping solution could achieve sufficient recoveries. Low grade recoveries are usually experienced.

Depending upon the concentration of the solder within the structure a reverse process can be adopted. Assuming that the solders as mentioned before were applied to nickel based materials. Such materials can be completely dissolved in warm HCL. This operation has to be well considered because of the quantities of material involved. However, almost all the materials can be recovered.

The residue after the dissolution of the structural material will contain pieces of unreacted silver solder. It will contain also gold in powdered form, silicon oxides, some of the alloying materials from stainless steel such as columbium, chromium and others. Great portion may be in silver chloride form.

This dissolution process may be the most prohibitive operation in the whole project, but it can be carried out in very simple tanks with little expense to the price of the chemicals.

The concentrated residue can be leached with aqua regia to dissolve a part of the gold content. This gold then can be precipitated after substantial dilution of the acid by addition of water. Sodium bisulphite is then added to precipitate the gold. The remaining residue of the silver solder contains palladium, gold and others. (Palladium can be also found in the liquid after precipitation of gold in the first leaching process, and also in the precipitate. It can be displaced from aqua regia by addition of zinc.) The residue is treated by the addition of sodium hydroxide in order to reduce chlorides into metallic form by heating the smelt to approximately 1800°F.

The resulting molten material will still contain all the elements dissolved in silver. This silver cast bars can be placed into silver refining tanks which will deposit pure silver and retain most of the remaining gold, palladium and columbium in the filter. This residue stripped of the mass of silver can be very well dissolved in aqua regia and then precipitated as in the original extraction. Processed gold precipitated out of the acid solution will still contain palladium as a co-precipitate. This of course, is removed in Wohlwill process by electrolysis where pure gold is deposited on the cathodes and palladium will remain in the electrolyte. Palladium can be recovered from the electrolyte. Combining the palladium residues from several steps of this process, we may re-dissolve the palladium in Nitric acid and precipitate it by sulphur dioxide. This is only an example of special projects which sometimes may be enterprised and the facilities should have provisions made for possible application for auxiliary processes and possible departures from regular activities.

SOURCES OF SCRAP

Material which has come in contact with gold, or contains gold, is a source of reclaimable gold. Many consumers of manufacturers have elaborate collecting systems for reclamation of gold scrap resulting from the manufacture of gold items and material which has come in contact with gold, such as grinding wheels, clothing, rags, and waste water. Elaborate precautions are taken in the jewelry and dental industries, primarily because of the high grade products and longevity of operations. Industrial users have varying degrees of sophistication in their scrap reclamation systems. Improved collection methods are being promoted by individual companies and precious metals refineries.

PRECIOUS METALS SEMI-REFINERS

The input to semi-refiners may be any kind of precious metal scrap, depending upon the company's specialization of product. Scrap is received in any condition and from any consumer sector, and is received on a toll basis. Methods of preparation range from simple melting and casting to more complex physical and chemical methods.

Output is bullion or residue (treated sweeps or other miscellaneous material) that is shipped to a refinery for recovery of god. Precious metals semi-refineries are smaller operations than refineries.

CONCENTRATORS

Industrial scrap containing small quantities of precious metals is readily available on the market for recycled materials. The technical capability to process this kind of materials depends primarily on the type of industry and the investment necessary to enterprise the recovery of all types of material. Even on the level of small industry it is possible to install a process for upgrading the waste material.

The processes for recovery of precious metals, however, becaomes a secondary objective of the business. The primary objective is to upgrade the base metal. Gold plated materials, containing more than 0.2% of gold is usually sold for the price of gold in the lot, and some allowance is made for associated materials, such as copper, silver, and others. The prices of these materials are quoted on the level of secondary commodity. These allowances are made only on open bidding and never offered to a direct seller.

The materials, containing small percentage of gold, but not contaminated by plastic materials are usually sold on the basis of copper II scrap materials and at very rare occasions some compensation is offered for traces of precious metals. For most of the cases no price is ever offered for electronic scrap which contains all kind of contaminants such as plastic or bakelite boards, lead solder, mixtures of aluminum covers and electronic chassis, iron, porcelain chips, etc. The only exception is the Kovar (low alloy steel), gold plated articles are priced good even when the concentration of gold is below the level of 0.15%. This is due to the ease of stripping of the gold in cyanide process. This material, if segregated in lots, can be stripped cleanly from its base metal and the metals resold.

The main concern of the concentrators is the value of the base material which is, of course, copper. The price of Copper II on the commodity market is for most of the time, half the price of Copper I which is seldom attained by resellers of scrap. Cathodic copper can be sold for Copper I price. Prices for copper offered to individual sellers fall to a fraction of Copper I prices.

The upgrading of the material is rather a scrap-yard operation. Hand sorting of the material, of course, is the first operation. The contaminated material is assorted into five types of material.

1. Pure copper - Electrical conductors, wires, and cables without insulating materials.

2. Pure copper - This material contains insulating materials which cannot be stripped by mechanical means.

3. Silver plated copper.

4. Electronic scrap - This scrap contains plastic boards and other combustible materials, however, is not contaminated by iron, aluminum or other base metals.

5. Scrap material - breakage - This material usually contains all kind of scrap materials, however it has to contain copper a a major part of the salvage.

The first two categories of material are not chemically processed. (If the combustions is not considered for chemical process). Number 2 material is stripped by combustion and then may be melted and cast.

The third category, which is the silver plated materials, is melted into anodes and processed.

The anodes containing silver are placed in the electrolyte, which is made up from copper sulphate and low percentage of free sulphuric acid. Silver is not soluble in this electrolyte and can be collected in anode bags, then melted into bars for refining of silver. The electrolytical process is operated at 0.15 to 0.25 Volts and 20-30 Amps per square foot of the cathode. The copper is deposited at the cathode while the silver remains in the anode bags.

The fourth category of electronic scrap material containing plastic and other contaminants is roasted at 590 °F., then the temperature is increased to 700 °F., and ultimately to 950°F., with access of air, until all the combustible materials are reduced to ashes and the copper metal for most of the load is converted to oxides.

The fifth grade of the scrap material is treated the same way as the material in the fourth category, however, the material is kept in a separate batch, depending upon the extent of the concentrating process.

In any case, the materials converted into oxides are screened for removal of large pieces, then moistened with water and roasted again, at 300°F. to complete the oxidation of copper to cupreous oxide (CuO).

The roasted material is then leached with hot sulphuric acid (10 to 15%) in lead lined kettles or tanks.

The copper and some of the other elements dissolved in sulphuric acid make up an electrolyte for copper depositing. Copper metals can also be removed from the solution by simply passing the solution over pieces of iron. Copper is cemented out of the solution and the remining traces of iron can be removed by washing the metal with dilute HCL. Copper is only slightly soluble in hydrochloric acid.

This can be done in wooden tanks specially constructed for this process. Cathodic copper prepared in this manner can pass for #1 copper, if resold to refinery.

The residue from the leaching process can be placed into a glass or porcelain boiling vessel and boiled in concentrated sulphuric acid to dissolve the silver and platinum group metals. The residue is gold which remains undissolved but still grossly contaminated by silica and other insoluble matter.

The sulphuric acid liquor can be siphoned off and passed into a tank of water and boiled. The silver can be precipitated on copper plates hung in the tank, copper (a less noble element), going into the solution.

The silver crystals are collected from the plates and can be melted into the bars.As subsequent refining of silver, which was prepared in this process, is desirable.

Gold, silver (which contains the platinum group elements) and copper concentrated in this manner are subject to purification processes and refining as it will be described in later chapters.

BY-PRODUCTS OF COPPER ORES

Very often the copper is not the only commercial product obtained from copper ores. Sometimes the by-products are of minor importance, but occasionally their importance may equal or surpass that of copper itself.

Many copper ores contain gold as well as silver. Platinum and the associated metals of platinum groups, palladium, osmium, iridium, ruthenium and rhodium are found in most copper ores which contain gold and silver. Often they are present in minute amounts, but they follow the gold and silver through the process and are eventually recovered when the gold and silver bullion is parted and refined.

Most gold-bearing copper ores contain only small amounts of gold, but such large tonages of ore are treated that the gold produced makes a substantial and valuable by-product of the mining. Domestic copper ores yield over ten per cent of the United States domestic gold production. Also, silver can account for 28% of domestic production. The platinum with the associated group was estimated up to 44 per cent of the world production originating from domestic ores of copper.

It is not the purpose of this book to describe the metallurgical process of the copper industry. It is important to introduce to the reader all conceivable sources of precious metals and clearly define the function of gold refining industry. The gold refining industry is always associated to other industries which have direct access to sources of gold. In the case of gold bearing ore or by-products of copper ore, refining of precious metals can be considered as a unique part of the copper industry.

The electrolytic refining of copper leaves about 0.5% to 3.0% of original anode weight in the residue, depending upon the purity of the copper anode. In addition to the elements, we have already mentioned, a certain amount of metallic copper enters the slimes. Part of this is in the form of small nodules which break off from cathode deposits, and part of it comes from the precipitation reaction in the solution of cupreous ions which exchange charges to form a cupric ion and neutral copper atom.

The treatment of these deposits varies in different refineries, depending upon the analysis of the material. In general, the residues are subjected to three basic operations.

1. Roasting - to convert copper metal to copper oxides, and then leaching with sulphuric acid to remove the copper.

2. The residue is subjected to series of oxidizing fusions, in order to oxidize the base metal imurities which pass into either slags or furnace fumes.

3. Parting process which is a part of the refining operations.

The following typical industrial flow diagram of some large industrial copper refinery indicate the complexity of the precious metals concentrating processes.

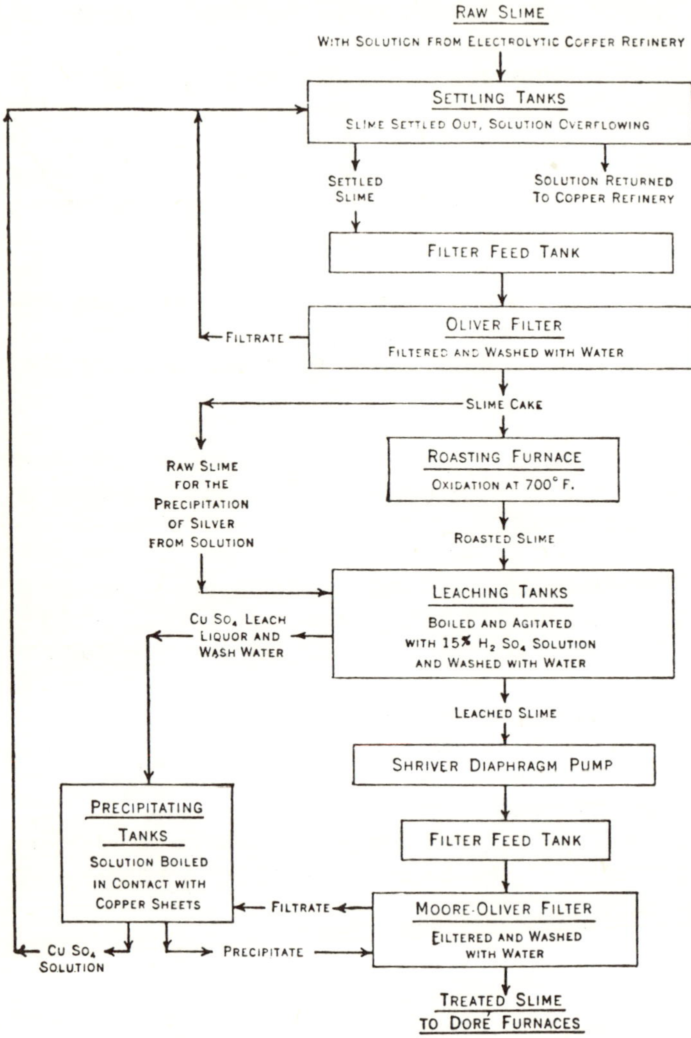

(Mosher, Am. Inst. Min. and Met. Eng. Trans., Vol. 106, p. 428, 1933)

Flowsheet of Treatment of Anode Mud or Slime.

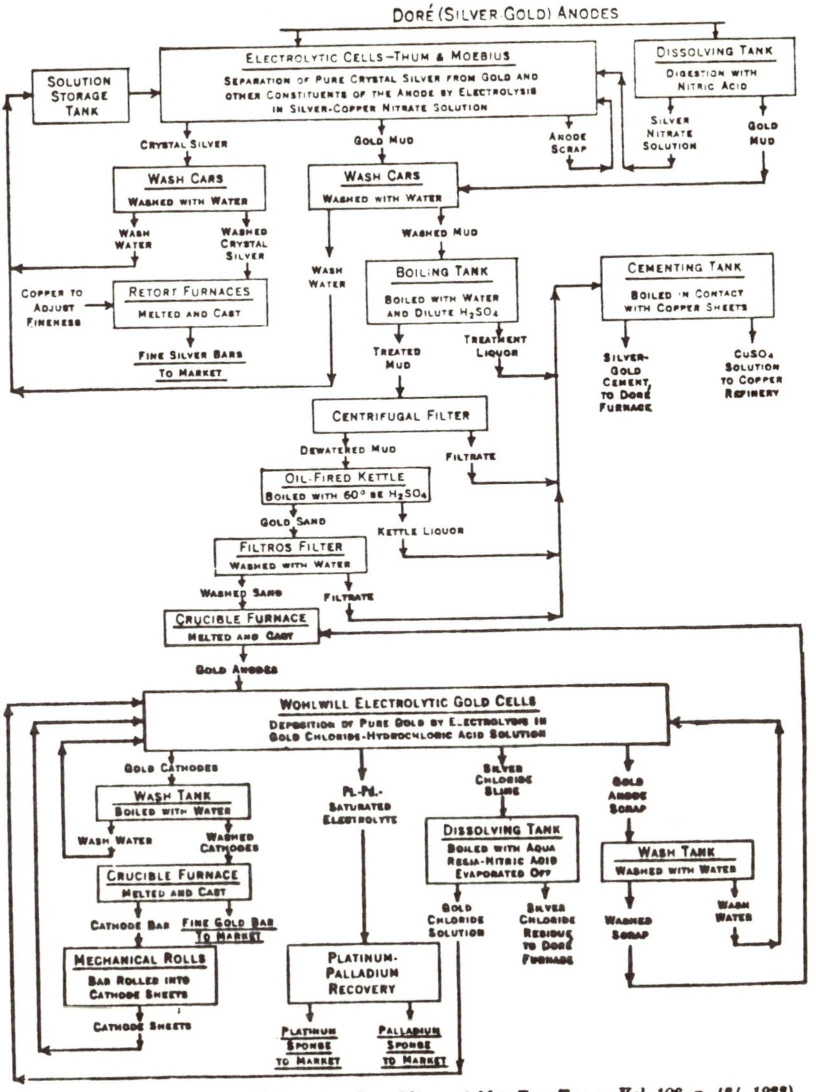

(*Mosher, Am. Inst. Min. and Met. Eng. Trans., Vol. 106, p. 434, 1933*)

Flowsheet of Electrolytic Parting and Refining of Precious Metals.

CONCENTRATES FROM GOLD MINING OPERATION

Concentrates of precious metals from mining operations may be products of a number of unique processes depending mainly on the type of ore or deposits from which they originate. The impurities associated with the gold will also depend on the origin, the type of ore dressing operations and chemical method which were applied.

Placer gold usually contains 65-82% of gold and a large volune of silicates, iron and other impurities.

Concentrates from mercury process will retain heavy metals.

Concentrates from gold smelters may contain large portions of lead, copper or silver metals.

In most of the cases, the concentrates can be handled in parting processes which are the purification step prior to refining of either of the metals.

PRIMARY BASE METAL SMELTERS AND REFINERS

Primary base metal smelters and refiners can handle as input any type of gold scrap from low-grade electronic to high-grade bullion. The gold content, associated metals, and other impurities present, determine where the scrap enters the recovery cycle. Recovery of gold from a base metal refinery is usually less costly than other refineries, but depends largely where the scrap enters the cycle.

If scrap goes through the entire recovery cycle, from material entering at the end of the cycle or in the electrolytic refinery section may require 2 weeks or less.

Input is received from consumers in the form of raw scrap; from semi-refiners in the form of bullion and residue; from refiners, and refiner-fabricators, usually in the form of bullion and residue; and from secondary nonferrous smelters in the form of impure base metals. Sources of the largest volume of high-gold content scrap are precious metal refiners.

CHAPTER FOUR

SEPARATIONS AND PURIFICATION OF PRECIOUS METALS CONCENTRATES

Concentrates of gold, or for this matter, of precious metals, may be of any nature or origin as it was described in previous chapter. The precious metals may be in a mixture of other elements even chemically separated from the gold but still in unresolved mass of compounds. Usually displacement of one of the precious metals will also cause a displacement of the rest of the metals in the group, if they are present in the solution. In cases when only one element of the group was present, a special technique must be applied to separate a clean portion of the material from the cementation process. In any case it is hardly a quantitative separation.

Precious metals, and also other base metals, must undergo several chemical separations prior to refining.

In this chapter it is intended to continue the description of the refining plant and processes as started in previous articles. The yield of the "in house" concentration process, the cyanide gold stripping operation, is a residue after zinc precipitation of gold from the alkaline cyanide solution.

The residue drained into a tow tray contains the following materials:

1. Diluted Zinc-cyanide solution.

2. Free Sodium cyanide solution.

3. Calcium hydroxide and Sodium hydroxide in solution and cyanide complexes of iron and others.

4. Gold in metallic precipitate.

5. Gold in solid aurous cyanide precipitate which is only soluble in excess of cyanide and slightly soluble in water.

6. Zinc metal in powder form (unreacted).

7. Insoluble compounds of silicon, calcium and other impurities.

The account of chemical composition of the concentrate serves only as a demonstration for understanding of steps in subsequent processing of the residue. It should be also emphasized that it originates from the simplest stripping action, on gold plated Kovar base articles, which were not contaminated with any other scrap materials. This material is considered as the easiest to handle in the plants. Any other batch of scrap will contain probably copper and silver metals.

As it can be noticed, it is desirable first to separate the liquid portion of the residue and replace it with water, so that very little of the cyanide compound will enter the next reaction.

1. The residue must be completely drained and

several times washed with water. (If nitric acid is used in subsequent reaction then distilled water should be used.) Trays are lifted on draining fixture and the aqueous part of the residue is siphoned to waste container. Then the washing is repeated allowing the residue to settle before draining. The washed residue is then drained, decanted into the boiling flask and the retained muddy residue in the tray is washed down by a stream of water.

2. The Pyrex glass boiling flasks are mounted in heating mantles which are movable on wheels. As described in previous chapter the flasks are wheeled to next processing facility which is physically separated by walls from the cyanide stripping facility. No contamination should be carried over to this space. The flask is placed under exhaust system and the electric heating elements are connected.

Any single acid which may be added to this mixture (in this case HCL is sufficient) will dissolve the residual gray mass of powdered zinc. Calcium or Sodium compounds are converted to chloride salts with evolution of hydrogen cyanide gas. This reaction is again accompanied by foaming and gassing and the addition of the acid must be carefully rated.

The residue is then boiled for several hours until the foaming ceases and a complete separation of the heavy metal from the liquid is obtained, then cooled to room temperature and the liquid decanted.

The residue in the flask now contains gold precipitated in metallic form, some insoluble silica and maybe other insoluble impurities. The remains of chlorides in the solution can be washed with water by decantion.

3. The gold residue from this process is sufficiently pure and free of interfering elements that it can be transferred to drying pan and then melted in the graphite-clay crucible. The insoluble impurities are removed by skimming the melt and then it can be cast into anodes for subsequent refining operation.

PARTING PROCESS

Metals in their elementary state and purity, exhibit special chemical and physical characteristics. For this matter all the elements can be assorted by these special properties into groups and sub-groups. This, of course, has been done very successfully in the past by Mendeliev and his unique work is a monument of achievement in the history of sciences.

The precious metals belong to two distinct groups of elements which exhibit very close chemical relationship and their chemical separation is affected in very particular ways. The chemical reactions in which gold, silver and copper completely separate is called a parting process, since its discovery.

It can be discerned that the presence of copper and silver in the concentrates of gold is a very common contamination or in reverse, copper ores will be for most of the time sources of silver and gold in minor quantities.

Separation of these elements depends on the quantity of each element in the composition. For example, a separation of silver from minor percentage of gold is different than separation of traces of silver from the bulk of gold concentrates. This rule in general is true in all the combinations of these elements and will be discussed in following paragraphs with reference to the processing in the refinery plant.

PROCESSING OF RESIDUE FROM CYANIDE STRIPPING

Residues which have been contaminated by Copper and Silver which inadvertantly originates from the substrate materials can be treated with dilute Nitric acid. The dilute nitric acid readily dissolves the zinc metal and also dissolves, on boiling, the copper and silver which is sometimes present in large quantities. This, of course, could not be achieved by hydrochloric acid in which acid both of these elements are insoluble.

The concentration of nitric acid should not exceed 25% by volume. For two reasons; the concentrated acid, unlike it is believed, dissolves gold, if the metal is in finely divided particles. A concentrated nitric acid will react violently with the metals and will decompose into nitric oxide gases. This can be avoided by dilution of the acid. In dilute nitric acid the metals react by substituting the hydrogen in the molecule of the acid. Boiling the residue for several hours in dilute nitric acid will result in dissolution of the metals without causing too much of polluting gases to escape. During this time, the reaction of copper and silver with the acid will cement out gold which could have been absorbed in the acid. After a separation of the residue is observed and the reaction ceases, plates of copper can be suspended in the boiling flask to ascertain that no gold is present in the solution. At the first sign of graysih deposit on the copper plates the cementation process can be discontinued. The grayish appearance of the copper surfaces are caused by crystalline deposition of silver and indicate the absence of gold in the solution. The solution then, can be decanted into a tank and cementation process continued as before. The residue in the flask consists now of high percentage of gold (99%). This residue can be washed with distilled water several times and then dried and melted into anodes for refining.

AN ALTERNATE METHOD FOR PARTING PROCESS

In case when there are no facilities for Wohlwill process and the chemical refining process is adapted, then a floowing process can be used.

The residue is dissolved in hydrochloric acid. The hydrochloric acid will dissolve primarily the excess of zinc metal in the residue and also convert the cyanide complexes into chlorides. This is accompanied by the evolution of hydrogen cyanide gas. After complete dissolution of the zinc, the liquid is decanted and discarded. To the remaining residue of gold, copper and silver, a dilute sulphuric acid is added. (The concentration of the acid should not be in excess of 25% by volume.) This acid will affect the dissolution of copper and other soluble impurities in the solution such as iron, stainless steel, aluminum and others. Silver is not affected. This solution also can be decanted and discarded. After removal of the bulk of the contaminating metals a concentrated sulphuric acid can be added and the solution brought up to boiling. The boiling has to be continued for a minimum of five hours to dissolve all the silver and remaining of the copper from the residue. The platinum metal group will also dissolve completely in this acid. The solution can be drained off to a tank of water which action will cause boiling of the sulphuric acid by dilution. The silver and platinum metal groups can be displaced from this solution by hanging plates of copper in the tank. The silver cemented out on the walls of the copper plates will contain the platinum metal group.

The dilute sulphuric acid can be stripped of copper by electrolysis and then re-pumped into the boilers, and boiled to sufficient concentration for use in the next parting process. Also the water can be collected for future processing so that there were no materials wasted. The facilities for this parting process and also the flow chart is attached in this chapter.

The chemical features of this process are extremely simple and rather of a physical nature. The sulphuric acid readily decomposes any complexes of cyanide and also dissolves most of the less noble metals in the solution. Silver in dilute solution is not affected and so is not the gold and other precious metals. In concentrated form, sulphuric acid does not attack any metal. When heated to boiling it will dissolve all the metals including silver to the last trace, with the exception of Gold. Gold can be removed after boiling in sulphuric acid in its very pure form, however, the washing of this acid is rather difficult.

ELECTROLYTICAL PARTING PROCESS

This process is specifically designed for stripping of heavy gold plates from copper articles. All the articles have to be electro-conductive and not contaminated with plastic materials. The plastic materials will prevent the continuity of the process.

This process is subject of a patent by Western Electric Co., Inc. Serial #87,693, Patent #2,185858 under the heading of "Method of Removing Gold, Silver, Palladium or the like"

The subject of the patent is to provide efficient method for removing the gold from copper, brass or bronze-base metals. This process also is efficient in removing gold from steel, stainless steel and other materials. The process uses sulphuric acid as an electrolyte. The specific gravity of the acid should be in the vicinity of 1.65. The sulphuric acid if it has sufficiently high concentration, above 60%, does not dissolve or attack most of the base materials which can be encountered in electronic gear. It leaves a protective oxide film on copper or brass articles and renders them completely passive during the operation.

The sulphuric acid electrolyte must be cooled during the operation at a constant rate to maintain the temperature below 100°F. Rise in temperature may cause a corrosive effect on the metal. When a current of at least 6 Volts is applied the sulphuric acid in the intimate vicinity of the anode forms a per-sulphuric acid in which the gold is dissolved and subsequently precipitates in the electrolyte in which the gold is not soluble. This anode layer is formed only on gold surfaces which display a substantially higher surface potential than copper, iron or silver, although silver is affected in similar way as gold in a much slower rate. A small amount of sodium trioxide or nickel sulphate was found to accelerate this action. This electrolytical process when applied to heavy plating can remove the gold from the substrate material leaving the bulk of the copper almost intact. Of course, on the long exposure of copper to the anodic corrosion the copper will also slightly dissolve in the acid. This process is very rapid for removal of the gold from the surfaces and quantitatively separating the elements of its group. The speed of the reaction is almost directly proportional to the surface potential of the metals.

This process can be carried for a long time in rather small quantity of electrolyte due to the fact that the reacting metals are being precipitated out of the solution. If an electrolytical tank is designed to store the accumulated precipitate, then large quantities of gold can be removed.

After the stripping action the electrolyte containing the cathodic mud, consisting of gold and silver, is transferred into a boiling flask. The adhering particles to walls of the tanks can be just washed down with a small stream of water. The boiling flask is placed into the processing stall and boiled for five hours. During the first two hours of boiling, sulphur dioxide is evolved which rapidly decomposes into pure sulphur. After sufficient boiling the silver which occludes the electrolyte is completely dissolved leaving on the bottom of the flask light brown-beige and rather pure gold precipitate. The electrolyte at this time is a completely transparent yellow liquid if no copper was dissolved. If copper is present, the electrolyte will have a greenish coloring. After a period of cooling, in which case up to 12 hours is necessary, the liquid can be siphoned off into a lead-lined dilution tank filled with water which volume is in the ratio of 10 to 1 to the electrolyte. The electrolyte entering the water causes a rapid heating of the water. For cementation of the silver metal out of this solution, copper plates are hung in the tank. The crystals of the silver can be scraped from the plates and melted into a cathode for purification process. The plates suspended in the liquid may be then connected to a direct current power supply as cathodes. By application of the current of 0.15 to 0.25 volts the copper, which dissolved from the plates into the solution cementing out the silver crystals, can be in this manner redeposited on the copper plates. Also, the sulphuric acid can be transferred into boiling kettle and boiled into its high concentration. This eliminates any waste liquids or electrolytes from the process. The gold residue is washed many times in water and the water is decanted into a holding tank. This gold is 995 pure and can be cast into cathodes for refining or it can also just be chemically refined as it is described in the next chapter.

THE PARTING PROCESS FACILITIES

The boiling of heavy acids in the flask requires a long time for heating of the liquids to boiling temperatures, and sustaining this action over a period of hours. It also requires a long time for the cooling off period. A typical operation of one batch of sulphuric acid parting may extend over 24 hours. The process can be accelerated only by installation of cooling fans in the heating mantles. This process does not require too much attendance. The important factor may be that the process can go on by itself without constant supervision, for this particular reason the facility must be correctly installed with the emphasis on safety precautions.

The Electrical Installation:

The fire hazards are eliminated by properly installed heavy electrical conduits in non-corrosive housings. A sufficient power supply should be provided. The power breakers should be installed outside of the room that they can be manipulated in case of an accidental breakage of the boiling flasks.

The Floors:

The cement floors should be free of any permanent fixtures or furniture. The walks between the boilers should be elevated an inch above the spaces for each individual boiling flask, which in turn are separated from each other by cement barriers. No floor drainage is necessary. The liquid loaded with precious metals, if spilled, should be contained in the space where they are located for processing. For the removal of the liquids from the floor, a special pulsating (self-priming) pump should be at hand. The pump is provided with hopolong hose (synthetic rubber hose) for sulphuric acid transfer.

The Heating Mantles:

The heating mantles are designed to allow the liquid to drain through the bottom of the heater without causing electric short and destruction of the equipment. The mantles should also be equipped with an air fan with a dispersing plate that the flask is cooled uniformly by whirlwind action.

Accidents:

The wall to the adjacent processing room is just a partition with glass windows that a constant observation of the facility is possible. In most of the accidents, the glass flask just cracks and the hot fuming liquid seeps out on the floor. In this case the electricity for the heaters is switched off and the room can be entered with a simple charcoal gas mask. The liquid is cooled by carbon dioxide fire extinguisher and then later re-pumped to an empty vessel. Explosions are very unlikely to happen, unless water is sprayed or drained into sulphuric acid; for this reason only carbon dioxide fire extinguishers are used.

Tanks:

A tank made out of Poly-Vinyl-Chloride (PVC) of 200 gallons is installed in this room. This tank is used for completion of the parting process. The boiling dilute sulphuric acid should not reach the temperature at which the tank would be damaged.

This tank can be used for nitric acid or sulphuric acid processes. The liquids are re-pumped from the boiling flasks by the same pulsating pump which may be used in emergencies. This pump is on a moveable support.

DRAINAGE OF CORROSIVE LIQUIDS

The facility should be provided with drainage of the corrosive liquids to the storage tank outside of the building.

CALIBRATION OF THE EQUIPMENT

Apart of the proper volumetric calibration of the process equipment a general perspective can be adapted by considering the basic characteristics of the process. A certain compatibility of the processing equipment must be determined at the time of the projection of the processing facility.

The most heavy and expensive equipment is usually necessary for the lowest grade of gold bearing materials. As an example, the mining operation requires, sometimes, dredging machines which constitutes a processing plant by itself. The low grade materials are voluminous and any handling requires substantial expense in help, storage, material handling equipment and large volumes of chemicals. The yield of gold and other precious metals is low per tonnage. The efficiency of the process is economically critical.

By each step of concentration of the materials, there is a step in magnitude. This step can be interpreted in smaller size of instrumentation. smaller and lighter handling equipment with diminishing volumes of liquids and weight of the materials. Ultimately, the precious metals are recovered pure in preciously small quantities.

For instance, there is only one boiling flask of 50-75 liters required for each cyanide precipitation tank of 200 gallons. Also, there are only 3 processing stalls needed for heating of these flasks at normal operation of the plant. With 6 cyanide precipitation tanks, only one furnace is required to melt all the gold which has been processed. Of course, for increasing versatility in processing of rather smaller batches the variety of equipment is more in order than the dimensions. Some of the equipment may be duplicated to be at hand for special purposes.

The same boiling flasks that are used for parting processes can be used for dissolution of jewelry and dental gold alloys. This means that there may be need for some extra boiling flasks just in case of accident or if the plant is very busy. The same facility can be used for dissolving the jewelry and dental alloys, however, special equipment and provisions are necessary.

EQUIPMENT FOR CONVERSION OF AQUA REGIA TO CHLORIDE SOLUTION

It has been described in analytical procedures, that conversion of aqua regia to chlorides can be ac- omplished by repeated evaporation of the liquids to moist crystals and then by repeated additions of the hydrochloric acid to the moist crystals, the nitric oxide fumes are expelled from the liquid and in this manner all the metals in the liquid are converted to chlorides. In processing of dental gold alloys, this conversion is very important in order to obtain a complete separation of platinum group from the gold and also for the recovery of each of the elements in platinum group. This procedure, however, is very lengthy and requires constant attention. If the chlorides which form crystals on the bottom of the vessel are accidentally overheated (250-300°F.) the chlorides of gold will volatilize and escape from the container. If unattended, it may cause a total loss of the gold. For this reason, a larger volume of the dissolved dental alloys is treated differently.

A larger quantity of the dental alloy is dissolved in 50 liter flask than the flask is provided with a specially designed condenser. The condenser is made out of pyrex glass tube, water cooled in outer jacket, the inner space of the condenser is provided with glass condensing tubes which inner surfaces can be drained into a collecting funnel. Collecting funnel reaches through the walls of the condenser and can lead off the separated liquid into another container. The diameter of the condenser matches the diameter (4") of the neck and height of either one of the processing flasks in the plant.

DISTILLING HEAD.

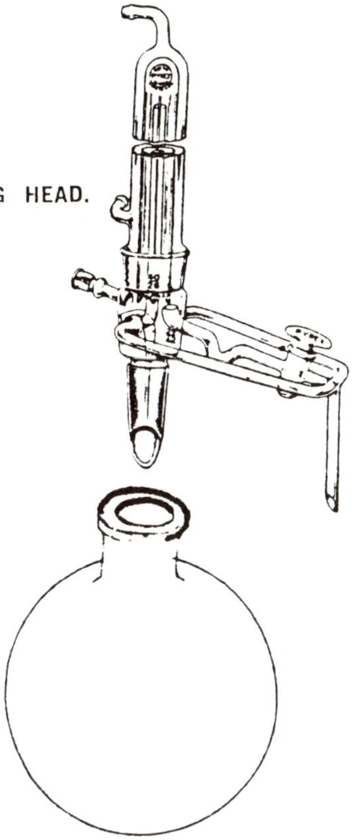

—This multiratio Corad (constant reflux ratio head) distilling head features a design incorporating the use of several condensing surfaces in parallel, the condensate from one surface being taken as product, while that from the other surfaces returns as reflux. It is fabricated with ratios of approximately 2.5:1, 5:1, 10:1, 20:1 and 30:1. These are marked on the condenser, which is attached to the lower part by a 40/26 joint, thus allowing rotation to any point desired.

Steam distillation may be carried out in the same manner as a dry distillation. The water and organic liquid, which condensate between a pair of guide lines, are delivered as product, while the rest of the mixture is returned as reflux.

The condenser is mounted on the wall on a moveable hinged type suspension which is loaded with springs so that easy handling is assured.

The boiling aqua regia in the flask is evaporated from the surface of the liquid and rises into the condenser. At first, it is condensed on all the surfaces and drained back into the liquid. In the process of cooling in the condenser, the liquid is giving off the heat to the surfaces of the condensing walls. The heat from the outer water cooled surfaces is lead off at a constant rate, while on the surfaces of the inner tubes the heat is accumulated until the fixture reaches the temperature of the boiling liquid.

There is a substantial difference of boiling temperature of hydrochloric acid and nitric acid. There is also a difference in boiling temperature of a mixture of these two acids. The boiling temperature of the nitric acid is higher than the one of the mixture and therefore, the nitric acid will condense on the surfaces of the inner tubes which are heated just to the temperature of the boiling mixture. The hydrochloric cannot be condensed on these surfaces because these surfaces are at a higher temperature than the boiling point of the acid. The hydrochloric acid and also the nitric acid will condense on the water cooled surfaces of the outer jacket and run back into the flask. The nitric acid which was condensed on the inner surfaces will drip off into the collecting funnel and is drained through the side of the condenser into the collecting vessel.

As the process continues, more of the nitric acid is stripped from the original mixture of the acids which will progressively boil at lower temperatures until it reaches the boiling point of the hydrochloric acid and remain at this constant temperature. At this

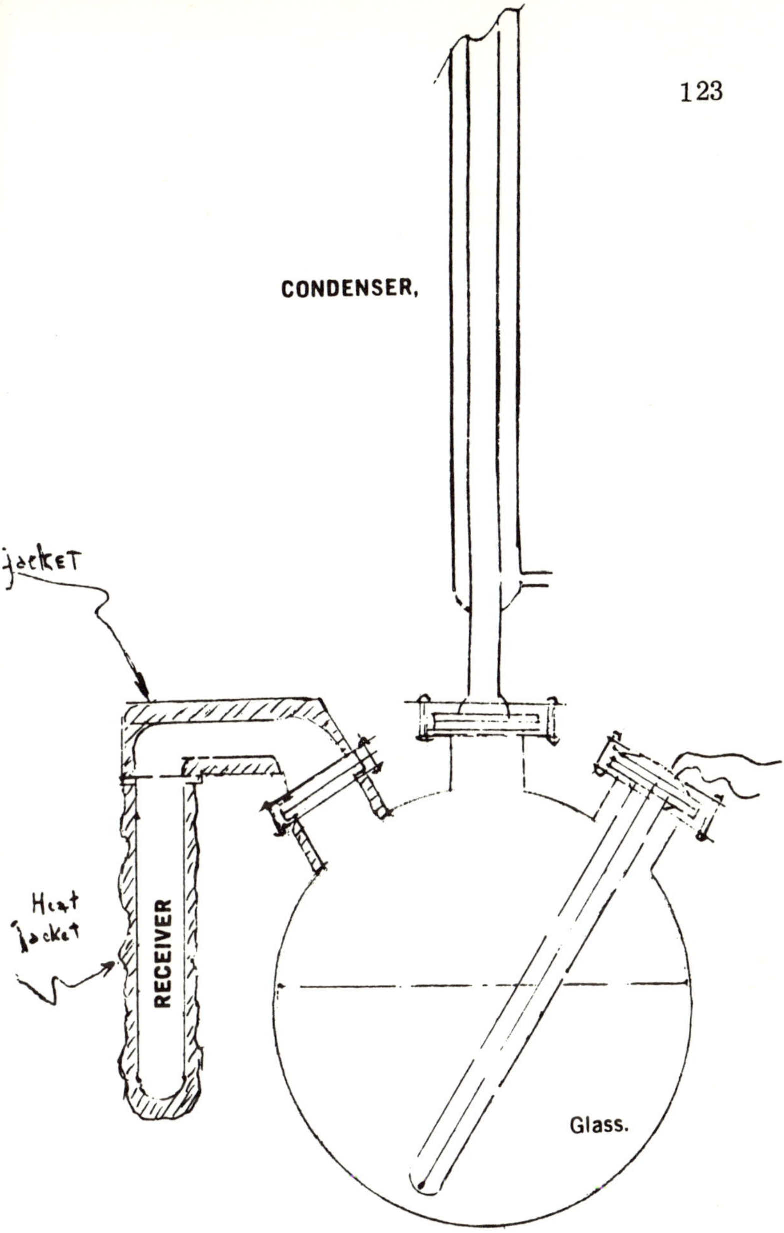

DISTILLING APPARATUS,

This distilling head is made in two parts. For use in collecting fractions during distillation time, there should not be any nitric acid present in the liquid and all the nitrates and oxichlorides in the solutions are decomposed to form chlorides. This process also expells gaseous phase of the decomposed nitrates. The gasses will escape through the open top of the condenser.

The separation of the acids is never completed at any stage of the process. The higher boiling phase is collected at the ration of 30% or sometimes higher, depending upon the location and height of the inner tubes and the heat applied. The collected higher boiling liquid is not nitric acid only, but still aqua regia of different ratio.

MELTING OF PRECIOUS METALS

Various techniques are employed in melting precious metals ranging from the use of 200 ton furnace in melting silver and rolling this silver into bus bars to the torch melting of small buttons. Besides these differences in the size of the charge, the differences in melting points and melting atmospheres and also the refractories used for different precious metals may influence the technique of melting.

POURING TEMPERATURES

The pouring temperatures depend on the alloy and on the pouring technique. It should be as slow as it will permit to pour all of the molten metal. Usually it is about 170°C. superheat above the liquidus state which is approximately 350°F. above the melting point of the alloy.

The temperature ranges from 650° to 1100°C. The dental alloys generally have the highest melting point and also contain most of the gold. The jewelry alloys melt at a lower temperature and contain less gold and alloys of much lower melting point.

The jewelry alloys including sterling silver contain copper as a hardener and require the same procedure as copper to secure ingots which are free of oxides. If cadmium is present then the melting of the alloy must be done under the fume hood because cadmium is volatile and very toxic.

For the purpose of melting the gold into the bars for Wohlwill process an electric furnace is sufficient.

The mantle of the electric furnace should be in a swivel suspension that the crucible could be tilted without being removed from the mantle. The whole furnace should be placed under the exhaust system for fumes which is built according to the city code for fire hazards. It should allow a full access to the crucible, however, the atmosphere above the crucible must not escape into the processing facilities. For control of the heat, a pyrometer within accuracy of ± 50°F. is sufficient. The total content of the crucible does not have to exceed 1 liter which means approximately 20 pounds of pure gold to 10 pounds of silver can be cast at a time. The crucible will never be filled completely to the top.

As it has been noticed, in the description of the processing of the gold from concentrates, with the exception of one filtration of cyanide solution, there were no other filtrations performed. This first step in gold melting serves several purposes. The melting of the concentrates after the parting treatment, and additional purification by melting with introduction of chlorine gas into the molten metal. The gassing of the melt with chlorine converts all the impurities which are present including copper and silver into chlorides which are for most of the time not volatile at this temperature. It also converts gold into chlorides which will escape from the crucible. At the first sign of red fumes escaping the surface of the molten metal, the chlorination is discontinued by pulling the ceramic tube out of the melt. The gold loss is negligible.

The melted metal will now contain chlorides of other elements which will float to the top of the melt, being of much lower specific gravity. With an addition of Sodium borate (Borax) flux, most of these impurities will be absorbed and can be skimmed off the top of the melt. Gold purified in this manner, can be poured by decantation from the crucible into open face molds for anodes which will be used in Wohlwill process. The size of the anodes should be compatible to the dimensions of electrolytical process which will be described in the chapter on refining. There should be a cast-in hook for attaching the anodes in the process. The shape of the anodes is conical narrowing down from the top to the bottom of the electrodes. This is because of the uneven corrosion of the anode experienced in the refining. The gold prepared in this manner will be 995 or better in fineness.

The same furnace will be used for melting scrap jewelry or dental alloys, for the purpose of dispersing the metal into very small articles. This serves the purpose of enlarging the contact area for the chemical reaction. If these alloys would not be dispersed in small particles there is a doubt they could be completely dissolved in aqua regia. This furnace should be provided with exchangeable crucibles for melting of different types of materials. The crucibles used for melting the jewelry and dental scraps should not be used for chlorination purification of the melts, or for melting of the silver.

INDUCTION FURNACE MELTING OF PRECIOUS METALS

The high frequency induction furnaces are being used increasingly for melting precious metals. The conductivity of these alloys is quite high and most of the units are fairly small so that relatively high frequencies are desirable if non-conductive crucibles are being used. Excessively high frequencies, however, cause too great heating for the metal where it is in contact with the crucible wall, due to the fact that higher frequency used, faster the surface is being heated. At lower frequencies, the heating zones are deeper in the metal, but the melting is slower accordingly. A proper frequency should be selected for the size of the crucible. Sometimes, it is to the advantage to use conductive carbon crucibles.

TORCH MELTING FOR PRODUCING BUTTONS

Special high silica crucibles cut off at an angle are employed for melting small quantities of precious metals. A torch is required that has been designed specifically for this method of melting. This melting is mostly used for melting small quantities of platinum or metals of platinum group due to the high melting point (3600 °F.) of these elements.

The torch is played on the charge until the metal is completely molten, forming a pool in the bottom of the crucible, then the flame is cut down slightly or is lifted away from time to time to permit the metal to solidify. The torch flame should be directed in such a way that solidification of the platinum will progress upward. If gasses evolve during cooling, the top of the button may be remelted to give a good surface.

After solidification is complete, the button is lifted out, and the clinging refractory is removed. The button is then placed bottom side up on a silica refractory and is rapidly heated just to the point of fusing the surface, smoothing it up, and melting any residual refractory.

Palladium should not be melted in this way, since there is a danger of contaminating it with silicon during the melting operation. For this reason, alumina lined crucibles are recommended. A strongly reducing flame of hydrogen is desirable and at the end of the melting period, the oxygen can still be cut lower to cool the flame and permit the progressive solidification.

SPECIAL PROCESSES

Jewelry alloys cannot be processed the same way without any filtration as it can be done concentrates from cyanide process. Jewelry alloys are not soluble in hydrochloric acid or nitric acid. Only a combination of these two can bring about the dissolution of the metal. It can be expected that the silver will combine with chlorine in this oxychloride solution and form an insoluble precipitate.

In reality, in concentrated aqua regia the silver chlorides are to some extent soluble. The aqua regia must be hot and must contain hydrochloric acid in the ratio of 5-1 to nitric acid. This ratio should be maintained during the dissolution by small additions of hydrochloric acid or additions of aqua regia with a ration of 1 nitric acid to 9 hydrochloric acid. This is necessary because there is a substantial amount of chlorine removed from the solution in the reaction due to the precipitation of silver chloride.

As the dissolution progresses the dissolved silver chloride will exceed its solubility in this media. To maintain this process of dissolution from the solid material will ultimately cause crusting of the silver chlorides on the surfaces of the metal and slowing the reaction. The completion of this reaction can be detected only when the reaction and foaming ceases.

This, of course, doesn't indicate that all the metal has been dissolved. Examination of the silver chloride precipitate and gentle crushing of the crust of the residue may give the idea when to discontinue this reaction.

PROCEDURE

The bulk of the metal is placed in a crucible and melted under the carbon powder cover in the furnace. No flux is used. The molten metal is slowly poured into a stainless steel container which is provided with a lid, with a hole in the center, through which the metal can be poured. In the center of the round container there is a flat stainless steel plate on which the metal is poured and from which this metal will be dispersed to all sides of the container and also rebound upward. On the bottom of the container there is water, but not reaching to the dispersing platform. The molten metal disperses into small beads which hit the side of the container and bounce back into the water and ultimately, solidify. To pour molten metal into the water, will not disperse the metal and the metal will settle at the bottom of the container fused in large chunks which would have to be remelted. This, of course, would not serve the purpose.

The same method is used for dental gold, except that the subsequent procedure for purifying the dental gold will be different.

The dispersed metal is collected from the container and screened to remove larger particles. The larger particles have to be remelted and dispersed again.

Finely divided particles of jewelry gold are placed into the boiling flasks and the aqua regia is gradually added until the metal is completely dissolved.

When the dissolution is finished, the solution is diluted with water to 5 times its volume. The dilution of this solution is also causing the liquid to cool off that re-pumping can be subsequently performed. The solution is re-pumped into a container for filtration and filtered through a #52 Watman paper of large diameter (24"). Sometimes, it is better that the solution is allowed to settle before filtration is attempted. If an immediate filtration is carried on, the first liter of the filtration must be returned to the filtrating funnel.

The filtered solution is then transferred (re-pumped) into a precipitation tank. Additional hot water at 140-160°F. is run into the solution to dilute it to its 10 to 1 volume or more. The solution at this time should be at least 110°F. of temperature. To this solution crystals of sodium-meta-bisulphite are added, approximately a pound per 10 oz. of the metal. This addition must be done under the fume hood or in a ventilated laboratory space. The sulphur dioxide gases escaping from the liquid are extremely obnoxious and poisonous. Of course, one doesn't have to be worried about smelling the sulphur dioxide gas, but it is rater important to avoid prolonged breating in a contaminated atmosphere.

The solution should be stirred during the addition of the crystal and then later covered for precipitation. This tank can be wheeled into a empty space in the lab and left to cool and precipitate for at least 24 hours.

After complete settling of the solution, the supernatant liquid is drained and discarded. The residue is washed down from the sides of the container and transferred into the suction filtration stand. The gold precipitate is filter-washed with distilled water until free of acid, using the suction to remove the hot wash water.

Then the residue can be dried, transferred into a crucible and melted into bars for refining. The residue is 999 pure.

If a large amount of copper is present in the alloy, there will be a substantial co-precipitation of copper. This will be removed by chlorination of the molten metal and by refining in Wohlwill process. The precipitate may appear dark due to the presence of copper and tellurium, which is usually closely associated with copper or gold ores. Also, there may be only traces of silver (in silver chloride form).

The residue from the first filtration will contain mostly silver chloride, some undissolved particles of metal, which contain gold and any insoluble material which was in the jewelry scrap. This can be transferred into a batch of other silver chloride remnants for processing of silver.

When a sufficient amount of silver chlorides is collected from this, or similar processes, they then can be dried and placed into the steel crucible and fused with sodium hydroxide. The sodium hydroxide will reduce the silver chloride to metal and the metal will absorb all the remaining particles of gold or other metals which may be present. After removing the upper layer of sodium-hydroxide and chlorides from the melt, the metal can be poured into a bar for silver refining process. In any case, there will be an excess of silver in the melt, that a parting of the silver is obtained in the silver electrolyte. Any material lost in the residue will be recovered from the residue in silver processing tank. This will be described in the next chapter.

Collected dental scrap must be freed of any ceramic or plastic contaminates at least to the degree that the batch can be considered a batch of metal scrap. The metal is treated by melting and dispersing in the water, in the same manner, as in the case of jewelry scrap, except that it will be done in a different crucible and different batch.

The reason is, that the remaining liquid after precipitation of gold will be treated for removal of platinum group metals, while in jewelry it can be simply discarded.

This scrap may contain even platinum alloys or pure platinum. The metal is dissolved in boiling flasks with aqua regia, following the same procedure as mention in jewelry scrap, however, a complete dissolution is not expected. The residue after this dissolution will be fused with potassium hydroxide and potassium nitrate 5-1, then dissolved with water. The metallic residue re-melted and re-dissolved in aqua regia. Both of the products can be added after filtration to the bulk of the solution.

The fusion is carried at 450° to 600°F. in a stainless steel crucible. The dissolved fusion in water is slightly acidified by hydrochloric acid and is added to the bulk of the dissolved sample. The solution prepared in the flask is rolled under condenser for conversion of the solution into chlorides.

Initially, there is a volume of acid (HCL) added to double the volume of the solution in the flask. The condenser is attached to the neck of the flask and the solution is converted into chlorides. This process does not have to be attended for some time. The exuded liquid from the solution should be saved and cooled for the next dissolution of metals.

When the solution attains the constant boiling point, at this time it will appear black-brown, iridescent. The process is discontinued and the solution is cooled down to room temperature. The volume is doubled by dilution with distilled water. The solution can be pump-filtered into another container. To the cold solution, ammonium chloride is added in liquid form (20% of aqueour solution), until no precipitate appears at the addition. The precipitate of ammonium-chloroplatinate is fluffy and bright yellow in appearance. It may appear to settle down at the bottom and at the same time float on the top of the liquid. The solution is allowed to stand over a period of hours. The solution is pump-filtered into the original flask from which the residue has been completely cleaned. The residue in that case contains silver chloride and can be treated in the same manner as jewelry scrap. The yellow precipitate of ammonium-chloroplatinate is completely dried.

Precipitation of the gold from the remaining solution is affected by the addition of ferrous sulphate (in aqueous solution) acidified by hydrochloric acid. It will require 4 1/2 times the weight of the expected gold from the dental scrap which is estimated 60-70% of the total weight of the batch. This means that if we dissolved 1 1/2 pounds of dispersed dental alloy in the original dissolution of material, then 4 pounds of this precipitant is the minimum which must be added to this solution to affect the precipitation of the gold.

The solution has to be heated almost to boiling to affect the precipitation. The gold will separate completely in a very pure form from the alloying material and the precipitate will be of light beige-brown color. The super-natant liquid containing palladium and maybe other elements is transferred by pump or siphoned off into another boiling flask. The gold residue will be washed several times with dilute hydrochloric acid (1%) and the wash is added to the liquid filtrate after the precipitation of gold.

Palladium in dental alloys may range between 2 to 20 percent of the weight of the alloy.

Palladium is precipitated from the remaining solution by the addition of dimethyl-glyoxime. The precipitated Palladium-dimethyl-glyoxime is a canary yellow fluffy compound which has the same physical characteristics as the corresponding nickel salt. It will separate at once from the solution and can be filtered out by coarse filter paper (#30 Watman), then washed with hot distilled water and dried at 220°F.

An addition of granulated zinc to the remaining solution after the palladium precipitation will displace all of the remaining precious metals of the platinum group if they were present in the alloy. For most of the time, dental alloys may not have other alloying constituents, however, presence of platinum and palladium in the alloy usually carries other elements of the group in small quantities. The precipitated metal, if any, is washed with hydrochloric acid to remove zinc and then stored for later reclamation after a sufficient amount of platinum group metals are accumulated.

TREATMENT OF THE PRECIPITATES FROM DENTAL ALLOY

The precipitates from this process are not in metallic form with the exception of gold.

GOLD: Gold precipitated by ferrous sulphate from chloride solution is in very pure form and can be melted into anodes for electrolytic refining. An effective precipitation of gold could be achieved also by addition of oxalic acid to the chloride solution of the metals. In this case the precipitation of gold could be carried as first reaction with subsequent separations of platinum and palladium.

If ferrous sulphate is used, then platinum metal must be separated first and then the gold metal, and palladium last. Large amounts of ferric salts will interfere with complete precipitation of platinum. Also, dimethyl-glyoxime will precipitate gold and palladium in chloride solution. The sequence of the operation must be observed.

PLATINUM: The precipitated Ammonium-choloro-platinate is a distinctly canary yellow compound. A greenish tint of the precipitate indicates the presence of Rhodium. Rhodium, even in small amounts in the solution will color the solution to a deep cherry red. If any quantity of ruthenium or other elements of the group is present, then the precipitate will be yellow-brown.

In any case, the dried precipitate is transferred to a porcelain dish and ignited at 900°F. to 1100°F. After volatilization of all of the ammonium salts, the precipitate will turn into a light grey mass of platinum metal sponge, and can be weighed as metal. This metal is 90% pure.

PALLADIUM: The dried palladium precipitate can be ignited at 750°F. and then in hydrogen flame using the reducing flame with little or no oxygen over the metal.

SILVER: Silver in form of silver chloride can be reduced by fusion with sodium hydroxide at 1800°F.

METALS OF PLATINUM GROUP

With the description of the special processes it becomes obvious that the conversion of aqua regia to chloride solution is important for dental scrap. Once the laboratory is equipped with this sort of process equipment, then the metals of the platinum group can be processed on a commercial basis. Not any special equipment is necessary for this processing because most of the separations are done by purification of segregated metals chemically, and in small batches. Maybe the only exception of the rule is the osmium and ruthenium which may be found in the alloys requires special distilling equipment.

Unlike in dental gold, in platinum metal alloys all the elements of the group can be found. Commercially all of the metals are used for some specific purposes, however, the largest marketing is in platinum and palladium. For iridium and ruthenium, the time of settlement may exceed six months. Osmium is mostly applied for manufacturing of ball point pen tips. Rhodium finds wide usage in electronics, and in plating for protective surfaces of delicate instruments such as analytical balances, standard weights, navigational devices and optics. Platinum-iridium alloys find use in petro-chemistry. Finely divided metals of these alloys are used as catalysts to accelerate the chemical reaction at lower temperatures. These metals are usually deposited on screens and devices with large surface areas. Scrapping of platinum-iridium screens is a very common occurrence.

Platinum alloys in dentistry are not as common as gold alloys, however in jewelry, platinum settings which are indistinguishable from white gold are frequent. With proper identification, these materials can be segregated into batches and treated for all the elements present. It is not essential to identify the alloys for the processing, however, it may be for estimating the price of the scrap. If the scrap is accepted, then the date of the settlement for the full price of the scrap is distributed for each element at a different time, so platinum can be settled within a period of one month, while iridium, ruthenium and osmium requires a minimum of six months for settlement.

A batch of scrap containing catalyst screens must be pre-washed in boiling hydrochloric acid to remove the remnants of the catalyst reaction.

The entire group of platinum metals can be brought into solution by using a mixture of barium peroxide and barium nitrate and extracting with dilute hydrochloric acid.

Hydrogen sulphite precipitates all metals from hot solutions with HCl content up to about 5 per cent. Iridium solutions should have about 20 per cent of HCl. The ignition of the sulphides yields metals in the case of platinum, oxidized residue with ruthenium, iridium, rhodium, palladium, and the volatile tetraoxide with osmium. Some sulphur will be retained in all cases in the residue. The volatilization of osmium is sometimes very desirable for the final purity of the platinum, however, if the osmium is needed to be collected then a special furnace is necessary for collection of the osmium. Furnace of this sort is usually an electric furnace with a quartz combustion tube 6-8 inches in diameter protruding through the mantle of heating elements. Both ends of the tube are capped with reduced entrance for gasses which are being passed over the metal, and with an exit into the absorption flask. During the combustion of sulphides, the air may be purged through the furnace and ultimately, the pure oxygen, the osmium will volatize and is trapped in alkaline solution fo sodium hydroxide. This furnace also can be used for chlolrination and reduction of the palladium in hydrogen atmosphere. If the refiner is not equipped with a furnace of this sort, then osmium can not be recovered.

When the metals are alloyed with lead, and dissolved in dilute nitric acid, some rhodium and platinum will go into the solution with the lead and by treating the wet residue with dilute aqua regia, the remaining platinum and rhodium will dissolve, leaving the iridium insoluble. These treatments provide a clean separation of iridium and rhodium.

IDENTIFICATION OF THE BINARY ALLOYS

In the recovery of the platinum metals, in the presence of other metals of the same group, it is almost inevitable to go through the complete separations for each element. To adapt this procedure is necessary because we really don't know what elements the scrap material contains. If the material could be identified and assorted in batches according to their alloys prior to the processing then the separations of the elements from the bulk of the materials could be simplified to one operation only. Most of the platinum alloys are binary alloys and contain only one alloying element of the group and platinum. The procedure could be chosen for a separation of a particular element and this could save considerable time and effort. The preliminary identification of the scrap material and the assortment of the material into batches is simpler than to process the material through all the stages. However, the methods of identification must be simple and rapid. Basically, all the platinum alloys, with some exceptions are binary alloys. (Binary alloys have only two elements present in the composition of the metal.)

The most common alloys in the industry are platinum alloys containing various percentages of rhodium, or iridium, or ruthenium and palladium. A dissolution of a sample by itself is a strong indication of the type of material which is being dissolved.

The platinum-palladium alloy or the platinum-ruthenium alloy and also platinum-palladium-ruthenium alloy are readily soluble in aqua regia, leaving no insoluble residue. A small sample of the metal can be dissolved in aqua regia and evaporated to moist crystals and then re-dissolved with hydrochloric acid. Platinum can be precipitated with ammonium chloride solution and separated from the liquid. If the precipitate is a bright yellow color the chances are that no ruthenium is present. An addition of potassium iodide, to the sample which contains palladium, will instantaneously precipitate black-palladous oxide, which can be re-dissolved by addition of excess of the reagent. If no precipitate appears at the initial addition, the sameple should be heated. After a time by heating, potassium iodide will precipitate a black sesqui-iodide of ruthenium, if the ruthenium is the element present.

Platinum rhodium alloys are more difficult to dissolve and an alloy containing 30% of rhodium is practically insoluble in aqua regia. However, both of the metals in the alloy will go into solution forming the chlorides of rhodium and platinum. The dissolution of the sample has to be continued with some addition of aqua regia until all of the metal is dissolved. A distinct cherry red coloration of the solution will indicate the presence of rhodium, even in a small amount.

Platinum-iridium alloys are virtually insoluble in aqua regia or only partially soluble, depending upon the percentage of iridium in the alloy. A solution can be affected with chlorine gas which yields chlorides or iridium and platinum. Solution of this sample is treated with ammonium chloride which precipitates black ammonium iridate and yellow platinum precipitate. The solution can also be tested addition of lead acetate which gives a grey-brown precipitate. If iridium is precipitated in presence of platinum, then it imparts pinkish to scarlet color.

PALLADIUM ALLOYS

Palladium is the only element of the platinum group which dissolves in nitric acid. Most of the alloys of palladium also dissolve in hot dilute nitric acid, even alloys containing gold. At least three times the weight of gold in silver should be present in the alloy in order to dissolve the palladium alloy in nitric acid, leaving the gold as the insoluble residue.

Materials which are identified by this simple testing of solubility can be assorted into batches which can be handlen in much more simplified way than it is indicated in the commercial separation of platinum group elements.

PROCESSING OF THE IDENTIFIED ALLOYS

Palladium alloys, which are soluble in nitric acid, are filtered by vacuum filtration and the residue is treated as gold (with dental alloys). The filtrate is diluted to four times its volume with water.

An addition of sodium chloride to the solution will precipitate all the silver in the form of silver chloride ($AgCl$).

Palladium is then precipitated from the filtered solution by addition of ammonium chloride crystals and heating the solution for short period of time in order to complete the precipitation.

The precipitated palladium is filtered from the solution and washed with alcohol and dried. The precipitate is then ignited at 1000 °F. The metal sponge is reduced to metallic palladium by moistening of the residue with formic acid.

Since the silver chloride tends to retain a substantial amount of palladium, it is advisable to redissolve the silver metal with nitric acid, after reduction of the silver chloride to metal, and re-precipitate as silver chloride to obtain a complete separation of palladium. This of course, is accomplished in silver electrolytical refining process. Palladium will stay in the electrolyte.

TREATMENT OF PLATINUM RHODIUM ALLOYS

If the major interest in the recovery of platinum is the resale of purified platinum then it is not necessary to adopt special procedures for each and every alloy in these separations because the platinum metal will be purified according to the next purification steps and then every metal will be affected in separation from platinum. An advantage can be gained by separating the platinum from its major alloys in one single process.

Platinum alloys, containing 3.5 to 20% of rhodium, but no other element such as iridium, are slowly dissolved in aqua regia. The aqua regia will have to be replaced from time to time as the alloy dissolves very slowly. After a complete solution of the alloys is accomplished the liquid is filtered off into a boiling flask. The aqua regia solution is converted into chlorides by refluxing. To the cold concentrated chloride solution of this alloy, a concentrated ammonium chloride is added which will precipitate all of the platinum leaving the rhodium in the deep cherry red colored solution. Even a very small amount of rhodium will discolor the yellow precipitate of ammonium chloro-platinate. The presence of rhodium in the precipitate is indicated by green tinge in the canary yellow precipitate. After thorough washing with solution of ammonium chloride and alcohol, the platinum is recovered from the precipitate by ignition at 1000-1100°F. This will cause the platinum to change into grey metallic sponge and later, it can be purified as it is described in the chapter on refining.

RECOVERY OF RHODIUM

The brownish-red filtrate is treated with sodium hydroxide until a yellow rhodium hydroxide begins to separate. The volume of the neutralized solution should be adjusted by dilution that the total content of rhodium does not exceed 10 grams per liter of the solution which is approximately 1 ounce per gallon. The rhodium is precipitated out of this solution by addition of barium chloride which is soluble in this solution and by subsequent addition of bariun carbonate and sodium carbonate. After the precipitation, the solution is heated to boiling. After boiling the solution for several minutes, the solution is allowed to cool and is filtered off. The remnants of the barium carbonate suspension can be washed from the walls of the flask with 2% solution of barium chloride.

The precipitate is placed into another boiling flask and completely redissolved with addition of 25% HCl. This solution should be again filtered after dilution, then heated to temperature just below boiling for 45 minutes. During this heating period, hydrogen sulphide gas is passed through the solution to precipitate the rhodium sulphide. The precipitated rhodium sulphide is filtered off at once. And the precipitate in the filter paper is washed with water containing a little of ammonium chloride. The precipitate is dried and ignited in the crucible. The ignited sulphides are reduced and cooled in stream of hydrogen, yielding the metallic rhodium.

PLATINUM-IRIDIUM ALLOYS

Platinum alloys containing 5-30% iridium are generally insoluble in aqua regia. These alloys are smelted with lead at 1832°F. The addition of lead to the metal should be about 10 times the weight of the alloy. A graphite crucible is suitable for this operation. The crucible should be tapered to facilitate the removal of the cooled ingot. After removal of the lead ingot from the crucible, the ingot should be brushed free of any remaining carbon. It is not advisable to pour the hot metal out of the crucible because most of the iridium which will settle to the bottom of the crucible, could not be removed.

To the metal in the boiling flask, add dilute nitric acid, approximately 25% by volume, then heat up the acid solution to approximately 170°F. and allow to dissolve the metal and digest for several hours.

Disintegration of the lead ingot is usually completed in about two hours, depending upon the size of the ingot. Dissolution of the ingot leaves a rather volumnious greysih-black mass in the boiling flask. The solution is diluted twice its volume and decanted out of the boiling flask, washing the flask with water. The solution is then filtered through suction filtration funnel using a fine filter at the bottom of a heavy pad of filter pulp which form a thick filtering pad in the funnel. The residue after this filtration should be washed thoroughly with hot distilled water in the suction filtration funnel.

The lead filtrate solution and washings may have to be filtered twice or until the liquid is clear. The residue from the second filtration is added to the residue from the first filtration and both placed into a boiling flask. The residue is then treated with aqua regia to dissolve all the platinum from the residue. The aqua regia is added to the residue in a certain manner. First, one liter of water is added for each two ounces of the alloy before fusion, then 300-320cc of concentrated HCl is added, and then finally, in this order, 50cc of nitric acid added (specific gravity 1.42).

This ratio of acid and the order of addition is important to affect the solution of platinum from the slowly dissolving residue. The solution of the platinum is affected at the temperature of 185 °F. The iridium is insoluble in aqua regia under these conditions and remains as an insoluble residue in form of fine crystals of high density and bright metallic luster. This solution must be filtrated to remove very fine crystals of iridium which are adhering to the sides of the glass. The glass walls may have to be policed by ashless paper to remove the remnants of the metal from the walls of the flask.

Wash the filtered iridum metal thoroughly with hot water and alternately with 1% of hot HCl and finally, with hot water only. The remaining solution should be filtered at least twice. The collected crystals of iridium are placed in a porcelain crucible with the remnants of the paper pulp and dried in the air before igniting, then ignite at 750°F. and raise the temperature to 900-1000°F. and ultimately to the red heat at 1350°F. After ignition in the air, allow the iridium to cool in atmosphere of hydrogen. This can be done in the following manner:

The hot crucible is placed under the fume hood. All the fire or heating elements are switched off. A hydrogen torch is ignited (with no oxygen gas) just to form a reddish-yellow flame which is played over the residue. By lowering the tip of the torch close to the residue the stream of cool hydrogen is passed over the metal before it is combusted in the air at the top of the crucible. When the initial redness of the residue disappears the hydrogen torch can be extinguished and the hydrogen valve reopened, thus allowing the cool gas to pass over the cooling residue. Care should be taken that no sparks or rubbing metal in the exhaust system re-ignites the hydorgen in the air. This of course, would cause an explosion. The same process of reduction can be done in an electric furnace, as it has been described, for use in distillation of osmium where first, all the air is displaced from the tube by nitrogen gas and then the hydrogen gas is passed over the hot residue and can be collected in the burner and ignited.

The recovery of the platinum from nitric acid lead solution is usually accomplished by addition of zinc to displace all noble metals out of the solution and then the residue is washed by HCl and subsequently treated in the same manner as it is described in commercial separation of the platinum group elements.

PLATINUM-RUTHENIUM-PALLADIUM ALLOYS

The platinum-ruthenium alloys and the platinum-ruthenium-palladium alloys are the easiest to dissolve with aqua regia as in comparison with rhodium and iridium-platinum alloys. These alloys usually pass into the solution without leaving any undissolved residue. The coloration of the solution is yellow and sometimes greenish which can be very well distinguished from rhodium or iridium alloys.

The platinum-ruthenium alloys are handled in the same way as the platinum-ruthenium-palladium alloys except, the separation of the platinum from these alloys, by precipitating the platinum with ammonium chloride is unsatisfactory when large quantities are handled. The platinum would remain grossly contaminated with ruthenium.

The assorted metal is placed into a boiling flask and dissolved by additions of aqua regia until the solution is affected. The solution is diluted with equal amount of HCl and converted to chlorides. If any silver precipitate appears it is not necessary to filter the solution at this stage. The solution is allowed to cool somewhat, and sulphur dioxide is passed through the solution to precipitate all the metals as sulphides. The precipitate is allowed to settle and the clear solution can be decanted. The rest of the solution with precipitate is filtered by suction filtration until all the liquid is separated from the residue. Sometimes a double filtration is necessary. The residue after washing is dried and placed in a porcelain or clay crucible and ignited and then roasted under the hood or in the combustion furnace to remove all the sulphur and reduce the residue to metallic form. The residue is transferred into a silver dish or crucible and fused with a mixture of 1 KNO_3-5 KOH, at 600°F.

The fusion product after two hours of heating is cooled to room temperature and carefully redissolved in water. Then the solution is transferred into distilling apparatus and chlorine gas is passed through the alkaline (orange colored) solution. The chlorine distills off the ruthenium tetro-oxide which can be passed over to the receivers. The receivers are containing solution (10-15%) of KOH and alcohol. This alkaline solution containing ruthenium tetro-oxide distillate is evaporated to a smaller volume and the ruthenium can be precipitated out of this solution on boiling with absolute alcohol.

An alternative precipitation of ruthenium from the alkaline solution can be achieved by acidifing the solution with HCl and precipitating the ruthenium with hydrogen sulphide gas then filtered, dried, ignited and reduced by hydrogen. Another alternative method uses dilute HCl in the receivers. This solution must be evaporated in a porcelain dish and the metal is reduced in baking. The metal is then reduced by hydrogen gas to metallic form. Also, an addition of formic acid to the HCl solution precipitates metallic ruthenium on boiling.

The remaining solution from distillation is acidified by hydrogen chloride and then filtered. The solution is treated with ammonium chloride to precipitate all of the platinum. Through all these sequences of operations, the solution is cold. A warm or hot solution would not allow a complete precipitation of platinum or palladium.

The precipitated platinum in the form of ammonium chloro-platinate is filtered out of the solution and thoroughly washed with a solution of ammonium chloride and alcohol in order to remove traces of palladium in the retained liquid in the precipitate. This wash solution can be added to the filtrate, however, a large quantity of alchohol may slow the precipitation of palladium.

To the filtrate a 1% of dimethyl-glyoxime in dilute alkaline solution of sodium hydroxide is added. This solution should be in proportion where 2.2 grams of solid reagent is required for every one gram of palladium.

The yellow palladium dimethyl-gloxime will precipitate out of the cold solution and can be removed by filtration through a coarse filter paper. The precipitate is dried, ignited and the metal reduced by hydrogen gas. This separation is complete and only one operation is necessary to remove all of the palladium from the solution. If gold contamination is suspected then the gold has to be removed from the solution prior to this separation. This can be done by the addition of ferrous sulphate or oxalic acid to the filtrate prior to the separation of the palladium.

The expense of the operation connected with dimethyl-glyoxime is rather substantial. An alternate and more economical method can be used for removal of palladium. Mercury cyanide $Hg(CN)_2$ is added to the nearly neutral chloride solution, to precipitate the yellowish-white palladium cyanide. This precipitate also can be filtered, dried, ignited and reduced to metallic palladium by hydrogen or it can be re-dissolved with HCl and the palladium can be reduced on boiling with formic acid to black metallic powder of palladium.

COMMERCIAL SEPARATION OF PRECIOUS METALS

A bullion containing gold, silver and platinum group metals, but free from base metals and osmiridium is digested with hot dilute aqua regia.

The solution will contain platinum, palladium, and gold. The iridium, ruthenium, rhodium, osmium, and silver will remain insoluble in the residue. The residue, after thorough washing, can be smelted with lead.

The smelted metals are dissolved and extracted with dilute nitric acid. The nitric acid extracts contain mainly lead and silver. The silver, of course, can be reduced from this solution by addition of sodium chloride and consequently precipitated as silver chloride. This is a product of the separation #1.

The remaining residue from this extraction contains Ir, Ru, Rh, Os. This residue can be fused with sodium bisulphate ($NaHSO_4$). This fusion product is dissolved in water. The aqueous solution of the sulphate will contain rhodium in sulphate form. This is a separation product #2.

The residue after this second extraction still contain iridium, ruthenium and osmium. After thorough washing and drying, this residue can be placed in a silver dish or crucible and fused in a mixture of one part of potassium nitrate (KNO_3) and five parts of potassium hydroxide (KOH). After fusion at 600°F, the crystals can be extracted with water. The ruthenium and osmium were fused in the salts and the insoluble residue now contains only oxidized iridium. The extraction product #3.

The extract is an alkaline solution of the salts and contains potassium ruthenate and potassium osmate.

A chlorine gas is passed through this alkaline solution and it will distill off both elements. The volatilized elements can be collected in dilute HCl or methyl alcohol. This solution can be evaporated to dryness. The residue is then placed in a furnace and heated to 1000°F. Oxygen gas is passed over the residue. Osmium oxide will be volatilized by the stream of oxygen and is absorbed in ethyl alcohol. The absorbed osmium oxides can be reduced by hydrogen gas to metallic state. This is product #4.

The residue after volatilization of osmium will consist of ruthenium tetra-oxides only. Moistening the residue with formic acid will reduce the ruthenium oxides to metal. This is extraction product #5.

The original solution from the extraction of the bullion contains platinum, palladium and gold. This solution is converted to chlorides. The gold can be precipitated out of this chloride solution by addition of oxalic acid. The reason why the gold is precipitated first is, that the oxalic acid will completely remove all the gold from the solution prior to the precipitation of platinum. Therefore, the platinum, which is subsequently precipitated from the solution, does not have to go through the purification process for removal of gold. If any platinum contamination remains in the gold, the contaminating material will be collected in the Wohlwill process. The precipitated gold metal is the extraction product #6.

The remaining solution is treated with addition of ammonium chloride which precipitates platinum in the form of ammonium chloro - platinate, and can

COMMERCIAL SEPARATION OF PRECIOUS METALS

- Bullion containing platinum metals, gold, and silver but free from base metals and osmiridium. Digest with aqua regia
  - Residue: Ir, Ru, Rh, Os, Ag. Smelt with lead and extract with nitric acid
    - Residue: Ir, Ru, Rh, Os. Fuse with $NaHSO_4$ and extract with water
      - Residue: Ir, Ru, Os. Fuse with KOH and KNO_3
        - Residue: Ir (oxidized). Add aqua regia to form $IrCl_4$ Add NH_4Cl Ppt.: Ignite → Ir
        - Solution: K_2RuO_4, K_2OsO_4 Add Cl_2 and distill
          - Distillate: RuO_4, OsO_4 Reduce with CH_3OH and HCl. Evaporate to $RuOCl_2$ and OsO_4. Heat to redness
            - Residue: Ru
            - Vapor: OsO_4 Reduce: Os
      - Solution: Rh
    - Solution: Pb, Ag
  - Solution: Pt, Pd, Au Add $FeSO_4$, ppting Au Add NH_4Cl
    - Ppt.: $(NH_4)_2PtCl_6$ Ignite → Pt
    - Solution: $PdCl_2$ Add $^{2}(NH_3$ $PdCl_2(NH_3)_2$ Ignite → Pd

and can be filtered off. This is an extraction product #7.

The remaining solution contains palladium and maybe remnants of each of the products. Palladium can be precipitated from this solution by di-methyl-glyoxime. This is a product #8.

The remaining of the solution is acidified with HCL and granulated zink metal is added which will displace all of the precious metals out of the solution. This is a product #9.

The residue with an excess of zinc metal is treated with hot hydrochloric acid. After the decantion of the zinc-acid solution, the residue, if any, can be recycled in the same process with another batch.

PRODUCTS:

1. Silver Chloride
2. Rhodium Sulphate
3. Iridium (oxidized)
4. Osmium Metal
5. Ruthenium tetra-oxide
6. Gold Metal
7. Ammonium-chloro-platinate
8. Palladium-di-methyl-glyoxime
9. Precious Metals Residues

The residual material from the separations which has been accumulated for reprocessing will probably contain all of the elements in the group. If sufficient amount of the residue is accumulated from prior separations, then the material can be subjected to the process again.

REFINING OF PLATINUM METAL GROUP

The production of chemically pure platinum is a difficult task, which involves the exclusion of the base metals and the elimination of the other members of the platinum group. The methods commonly employed for this purpose are similar to those used in chemical separations as it has been outlined in the previous chapter.

The electrolytic deposition of the pure platinum metals is giving some difficulty because it is impossible to use these metals as soluble anodes for continuous electrolysis. As a result, a platinum chloride bath has a short life and special techniques must be used to replenish the bath as the metal is plated out. In presence of other metals than platinum in the electrolyte, these metals are deposited first, leaving the platinum in solution. After all, these are the principles upon which the electrolytic gold refining process is based. In Wohlwill process the gold is deposited on a cathode leaving the metals of platinum group in the solution.

CHEMICAL REFINING OF PALLADIUM

The products of the commercial separation of the precious metals are not expected to be pure. In the first step of the separation, where the bullion containing platinum metals and gold is treated with hot dilute aqua regia, it is indicated that the extract will contain primarily platinum, palladium and gold. Silver is completely excluded, although it may be present in the residue, because of its virtual insolubility in aqua regia. The metals of the platinum group are not completely insoluble and will possibly contaminate this extraction. From the following listing of precious metals in the bullion, which is free of other base metals, it may be obvious what contaminations we can expect in the first extraction of the metals.

Gold	Soluble
Silver	Insoluble
Platinum	Soluble
Palladium	Soluble
Rhodium	Insoluble, except if it originates from an alloy.
Ruthenium	Partially soluble, (dissolves very slowly).
Osmium	Soluble, yielding osmium tetro-oxide.
Iridium	Insoluble

Is is evident from the first extraction in aqua regia, that there are element which will partially be carried over with the extract and if not eliminated they will contaminate the yield of the extraction.

Gold can be reduced from this solution in rather pure form and moreover, it will be refined in the Wohlwill process. If the gold should be contaminated by any metals of the platinum group, then these will be collected in the electrolyte of the Wohlwill process.

Silver cannot be expected to be deposited with any of these metals especially after conversion of the solution to chlorides and if any silver chlorides are formed they are eliminated by filtration of the extract prior to the metal separations.

Palladium can be completely extracted from the

chloride solution in a very pure form. This separation is complete and a rather pure palladium di-methyl-glyoxime is obtained. This precipitate can be readily separated from the liquid, dried and ignited. Ignition of the of the metal in the air will eliminate the possibility of osmium contamination. The palladium can be reduced to metal by hydrogen gas, or by moistening with forming acid.

PLATINUM

As it can be seen from the table of solubilities of precious metals in the dilute aqua regia, some of the element like ruthenium, rhodium and osmium will be carried over in this extraction, leaving the bulk of these metals in the residue.

Ruthenium dissolves slowly in aqua regia and will contaminate the platinum, when precipitated with ammonium chloride. Also, the osmium can be found in abundance in this extraction.

Rhodium is insoluble in dilute aqua regia but if it originates from some of its alloys, it will precipitate with platinum. The presence of these contaminations will persist through the precipitation by ammonium chloride and through the washing of the precipitate with dilute ammonium chloride and alcohol.

Platinum is precipitated from this chloride solution by an addition of concentrated ammonium chloride and alcohol solution. The yellow precipitate of ammonium-chloro-platinate is separated from the liquid by filtration. A thorough washing of this precipitate with 1per cent of ammonium chloride solution containing 20% methyl alcohol is essential. After complete drying of the precipitate, the precipitate is placed into a porcelain dish and ignited at 750 °F. Then at 900 °F and ultimately at 1100 °F, until it is completely converted to a grey sponge metal. Then the oxygen gas is passed over the hot metal which will volatilize all of the osmium tetro-oxides. The volatilized oxides can be collected in alcohol solution or other media. The metal sponge after this treatment is completely redissolved in aqua regia and the solution converted into chlorides. The solution is then evaporated to volatilize the excess of acid and then the crystals or the syrupy solution diluted with distilled water. To this solution, sodium bromate is added (10% of aqeous solution) and the solution is boiled for at least 10 minutes, then neutralized with sodium bicarbonate to Ph of 7.5 to 8.0. At this time, the solution will darken by formation of a precipitate. Then a freshly prepared barium oxide is added in small portions. The rhodium and any other element in the group will precipitate on boiling and can be filtered off.

The filtrate which contains pure platinum is acidified with HCl and the platinum can be reduced from this hot solution with an addition of formic acid The addition of formic acid must be carefully done to prevent excessive reaction. The formic acid will reduce the platinum to black powder which can be considered a pure metal.

The reactions of metals of the platinum group with an ammonium chloride in chloride solution:

Platinum	Yellow precipitate
Gold	No precipitate
Palladium	No precipitate
Ruthenium	Brown precipitate
Rhodium	No precipitate
Osmium	Red precipitate
Iridium	Black precipitate

The residues from the first step after smelting with lead are extracted with nitric acid. None of the precious metals except palladium and silver is soluble in nitric acid. The silver can be reduced from this solution by addition of sodium chloride and also palladium if still present, can be removed in the second step by precipitation with ammonium chloride after the removal of silver. This treatment with nitric acid extraction eliminates these two elements from the subsequent reaction. The residue from the nitric acid extraction is thoroughly washed and placed into porcelain crucible and fused with sodium bisulphate. The sodium bisulphate will dissolve the bulk of the rhodium metal from the residue.

The reactions of platinum group in fusion with potassium bisulphate:

Platinum	Slightly attacked (Not present in the residue).
Gold	No action (Not present in the residue).
Palladium	Dissolves (Not present in the residue).
Ruthenium	No action
Rhodium	Forms soluble salt $K_6Rh_2(SO_4)_6$
Osmium	Partly oxidized to OsO_4
Iridium	Oxidized but not dissolved.

This fusion gives a very good separation of rhodium from iridium and the rest of the elements of the group. The rhodium can be washed out from this fusion by distilled water. The solution is neutralized and the rhodium can be precipitated by an addition of barium hydroxide. The precipitate which is obtained, can be heated to convert the metal to oxides and then reduced to metallic form by hydrogen gas.

The next fusion of the residue of iridium, ruthenium and osmium is done in the mixture of one part of potassium nitrate and five parts of potassium hydroxide. The fusion is done by heating the residue with salts in the silver dish or crucible at 450-600°F.

The fusion reaction with Platinum group:

Platinum	Attacked (Not present in the residue).
Gold	Attacked (Not present in the residue).
Palladium	Oxidized (Not present in the residue).
Rhodium	Oxidized (Not present in the residue).
Ruthenium	Forms a green salt which is soluble in water.
Osmium	Dissolves, forming K_2OsO_4.
Iridium	Oxidized to soluble and also insoluble iridates.

This fusion product is dissolved with water, placed into a distilling apparatus and purged by chlorine gas to distill off ruthenium and osmium. The ruthenium can be completely distilled off and collected with osmium in the hydrochloric acid. The metals can be reduced to metllic form by addition of alcohol or formic acid and boiled to black metallic powders. This metallic powder is completely washed, dried and placed into the furnace in order to distill off the osmium at 1000°F, by passing the oxygen over the heated metal powder. The residue will be pure ruthenium and the distillate can be later reduced to pure metallic osmium by boiling the distillate with ethyl alcohol in presence of hydrogen gas.

IRIDIUM

The residue from the chlorine distillation contains iridium, and some osmium. By an addition of aqua regia to the residue, the iridium will form chlorides ($IrCl_4$). A subsequent addition of ammonium chloride to this solution will precipitate

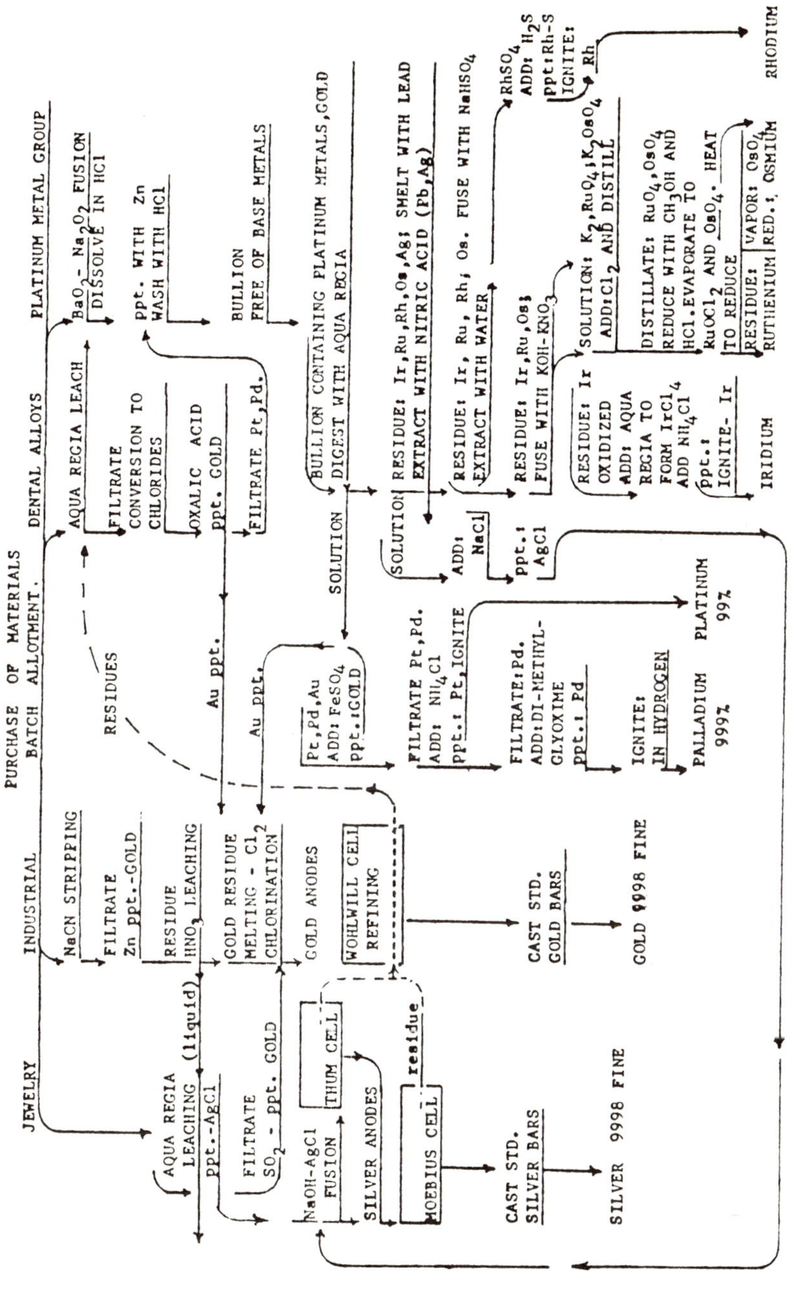

FLOWSHEET FOR PRECIOUS METALS REFINING

black chloro-iridate $(NH_4)_2IrCl_6$. This precipitate still contaminated by osmium can be converted to metal powder by ignition at 900 to 1000°F. and then by passing oxygen over the heated iridium, the osmium will be distilled off yielding pure metal.

CHAPTER FIVE

REFINING OF PRECIOUS METALS

The refining operations can be done only on those products which were treated chemically and through this processing separated from their alloying elements. The metal then is chemically purified and concentrated to the highest degree that the chemical system could produce. The raw concentrates as they may be obtained in the first steps of the processing must be first chemically purified before refining as they could destroy the refining process completely having still a large quantity of contaminates incompatible with the refining process. In some cases, a refining operation cannot be performed, due to the fact that a large quantity of remaining contaminates could displace all of the gold from the electrolyte. The impurities can be present only in very small quantities.

In the preceeding operations, the emphasis was placed on quantitative separation of the major elements in the composition of the material. It was important to extract all the gold completely from the contaminating materials, even at the price of carrying the contaminates along in the process. In the refining process the emphasis is on extracting the contaminates from the final product even if some portion of the main product stays in the processing solutions. These remnants are usually returned to the bulk of the material in the next processing batch, while the bulk of the materials from which the gold was extracted in the initial operations is usually returned to the scrap heap.

In these processes no loss of the precious metals is expected.

Gold and silver refining methods are two electrolytic processes which require a special facility which is separated from other chemical activities in the plant. In these processes a " crude " material in the anode is electrolyzed and a pure metal of high quality is deposited on the cathode, leaving the contaminates collected in the bags as insoluble matter or dissolved in the electrolyte.

SILVER REFINING PROCESS

This process is using a Moebius cell. In comparison to Thum cell, which is also used for silver refining and other kind of operations,these two cells are using basically the same electrolytic process except that the cells are arranged differently. The Moebius cell operates only with cast electrodes suspended from the bus bars, while the Thum cell can absorb the material in any form. The Thum cell can process contaminated crystalline residues and extract silver from remnants of other processes while the cast anodes for the Moebius cell must be of high grade of silver produced in prior separations. The silver from the Thum cell is usually refined in the Moebius cell.

MOEBIUS CELL FOR SILVER REFINING

The Moebius cell resembles a small multiple refining tank. The anode and cathodes are hung from suspension bars. The anodes, which are about 14" long and 5" in width and about half of an inch thick, are hung in canvas bags that are dense enough to hold the anode debris and slimes. The cathodes are plates of stainless steel or silver. The cathode silver deposits from the silver electrolyte form a loosely adherent crystals which are knocked off by scrapers during the operation, and then collected in a basket on the bottom of the cell. The cell dimensions are 24" by 26" with a depth of 22 inches. These tanks are constructed from poly-propylene or PVC materials. A filter basket is placed on the bottom of the tank, covering all of the bottom. The poly-propylene basket is covered with a glass cloth strong enough to be lifted from the bottom of the cell filled with silver crystals.

There are three rows of anode hangers with three silver bars suspended from each hanger. The anodes are provided with anode bags for each row. Five stainless steel cathodes are placed between the rows of the anodes on each side. These cathodes are provided with movable scrapers.

The cell operates with current density of about 50 amperes per square foot at a voltage drop 2.7 volts in each cell.

The electrolyte is a practically neutral solution of silver nitrates containing about 60 grams of silver and 30 grams of copper metal per liter.

Part of the electrolyte is removed each day and replaced with fresh electrolyte in order to keep the impurities below the limits and to build up the silver content as the electrolyte may become gradually depleted in silver. Every day at least once, the content of silver in the electrolyte is determined. The analytical method for determination of silver was described in Chapter Two #5 - a. When the electrolyte becomes very dark in appearance, it may be necessary to replace it. The foul electrolyte may be added to the Thum cell for depletion. The depleted electrolyte then can be diluted to twice its volume with water and the silver can be precipitated by addition of sodium chloride. A new electrolyte is prepared by using the pure silver from the Mobius cell. The silver is dissolved in nitric acid and the solution then boiled to concentration. The concentrated solution is allowed to cool and crystalie. Then re-heated again just above the boiling point of nitric acid and the excess of the nitric acid is evaporated. The vapor and the fumes of the acid are allowed to be carried away by a rapid stream of air. The crystals of formed silver nitrate are again re-dissolved with distilled water and the volume adjusted to such a concentration that it can be stored. The silver content of this solution is determined and then it is used for daily re-plenishing of the silver content in the electrolyte or it can be used to prepare a new electrolyte if it is desired.

Two pases of silver metal through this process will produce pure silver bullion 1000 fine.

THE SILVER EXTRACTIONS IN THUM CELL

This cell is a shallow tank about 50" by 24" and 9" deep. The tank is constructed out of polypropylene or PVC materials. The bottom of the tank is covered with stainless steel or graphite which serves as a cathode. A basket built in form of a shelf is placed on the top of the cathode. The perforated bottom of the basket is covered with a glass cloth, strong enough to support the metal. Approximately 3 to 3.5 volts per cell is required. The cathode is connected to the source unit by partial stainless steel and copper bus bars, while the anode (metal placed in the basket) is connected to the copper bus bar by a large solid block of silver metal, which serves as a contact between the bus bar and the scrap metal.

Although the operation and the purpose is the same,

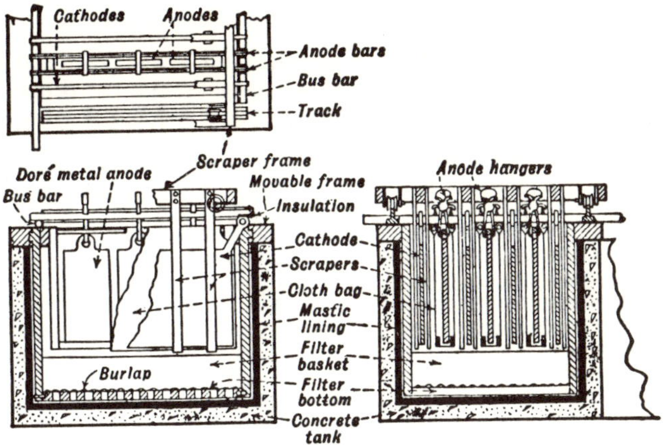

(*From Creighton and Koehler, Electrochemistry, John Wiley and Sons, Inc., New York*)

Moebius Cell for Silver Refining.

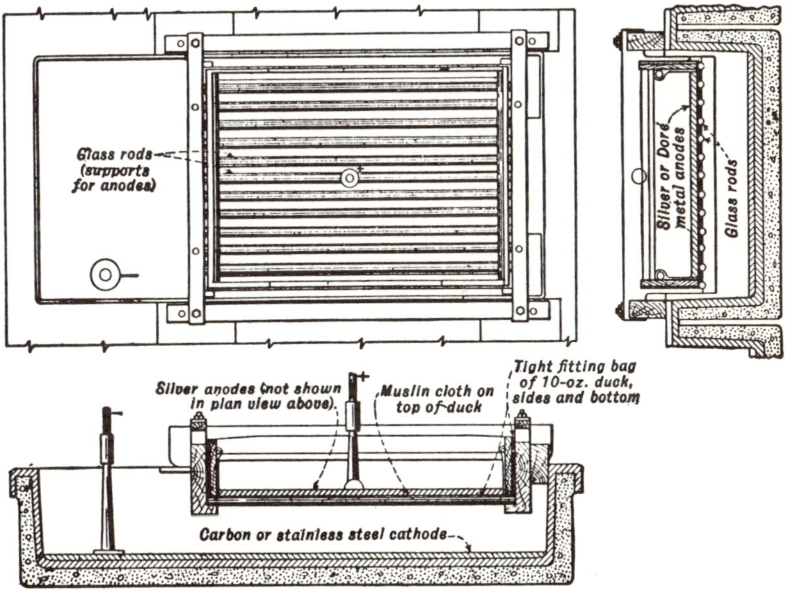

(*From Creighton and Koehler, Electrochemistry, John Wiley and Sons, Inc., New York*)

Thum Cell for Silver Refining.

this cell is mainly used for electrolytical parting of silver from crystalline residues, which can be conveniently placed into the basket. The silver deposited on the cathodes is scraped by hand scrapers. The access to the cathodes is facilitated by shifting the basket from one side of the tank to the other side. This cell produces less of scrap anodes because any additional material can be placed on the top of the basket while the remaining material is being eroded.

The anode residues from either one of the tanks can be combined and processed for extraction of precious metals. The silver produced in the Thum cell is reprocessed in the Moebius cell which renders the final silver product. The products of Moebius cell can be sold in crystalline form or they can be cast into 100 ounce bars each and sold as a commodity on the market.

RECOVERY OF PRECIOUS METALS
From the Residues of Silver Refining Process

The moebius cells produce much larger amounts of anode scrap than the Thum cell. If the anodes are not cast in proper shape they may even break off at surface level of the electrolyte leaving a substantial part of the anode in the bag. This scrap anodes are usually recast for recycling. The anodes should be cast into conical shape, suspended with the larger portion of the anode at the anode bars so that the erosion can gradually progress from the bottom up, leaving as little scrap anode as possible.

The combined residue from these tanks (except the scrap anodes), is thoroughly washed with distilled water. The remainder of the solid residue is placed into the boiling flask and dissolved in aqua regia and treated for chemical separations of all the precious metals as described in previous chapters.

THE GOLD REFINING
The Wohlwill Process

This process consist in electrolyzing impure gold anodes in a hot acid solution of gold chloride. The anodes are made from such materials which have been processed in the plant and purified to a minimum of 98% of gold. The gold is deposited on cathodes which are made out of rolled strips of pure gold and the resultant cathode gold will be from 9995 to 99995.

The cells are made of glazed porcelain and are small in size, because of the value of the electrolyte. Anodes and cathodes are suspended in the solution. The anodes are provided with glass cloth bags dense enough not to allow any particle of anode to escape. The electrolyte contains 7 to 8 per cent gold as gold chloride ($AuCl_3$) and 10 to 16 percent of free HCl. The electrolyte is maintained at a temperature of 160°F. High current densities (110 to 120 amperes per square foot) are employed to obtain rapid deposition of the metal. About 1.3 to 1.5 volts are required to give this current.

The Wohlwill process utilizes an alternating current superimposed on the direct current used for the actual plating when the silver content of the anode is high. This serves the purpose to prevent the silver chloride slime from adhering to the anodes. Same can be achieved by a low voltage rectifier which output is not filtered to remove the ripples in the current.

Part of the electrolyte is removed daily to control the impurities, and pure gold chloride is added to replace the gold lost by depletion of the solution.

Platinum and palladium concentrate is in the electrolyte, but some gold, silver chloride and other metals of the platinum group can be found in the anode slimes collected in the bag. These metals can be recovered as it was described in the previous chapter.

All three of these processes are used by the United States Mint and by commercial refiners. The material processed in this manner can be sold as 1000 fine and hallmarked.

THE WOHLWILL CELL

The electrolytical cell is constructed differently than the Moebius or Thum electrolytical tanks. This is because of the relatively high temperature at which the cell operates and the high efficiency in terms of output.

The cell is a 1.2 liter porcelain vessel with approximate dimension of 4" x 4" base, and approximatley 4.5" high. This little square porcelain tank is provided with a lip so that it can be suspended into a heating mantel. The heating mantel has radiant heating elements mounted at some distance from the wall of the tank. A thermostatic control operates at a temperature of 158°F. within the accuracy of plus or minus 1°F. A matching frame made out of glass or strong epoxy resin is placed on the top of the tank. This fixture has sealed bus bars protruding through the sides of the frame and a basket for the anode bag. The whole fixture can be lifted from the tank with the electrodes attached to the bus bars.

The top of the cell is provided with a lid (plenum) large enough to allow vapors to be condensed and run back into the cell. This feature is very important for a continuous unattended operation at elevated temperatures. This is because of the very small volume of the electrolyte and corrosivity of the vapor.

The cell is by no means sealed. The vapor pressure, which is relatively small, is constantly relieved by the condensation of the liquid on the

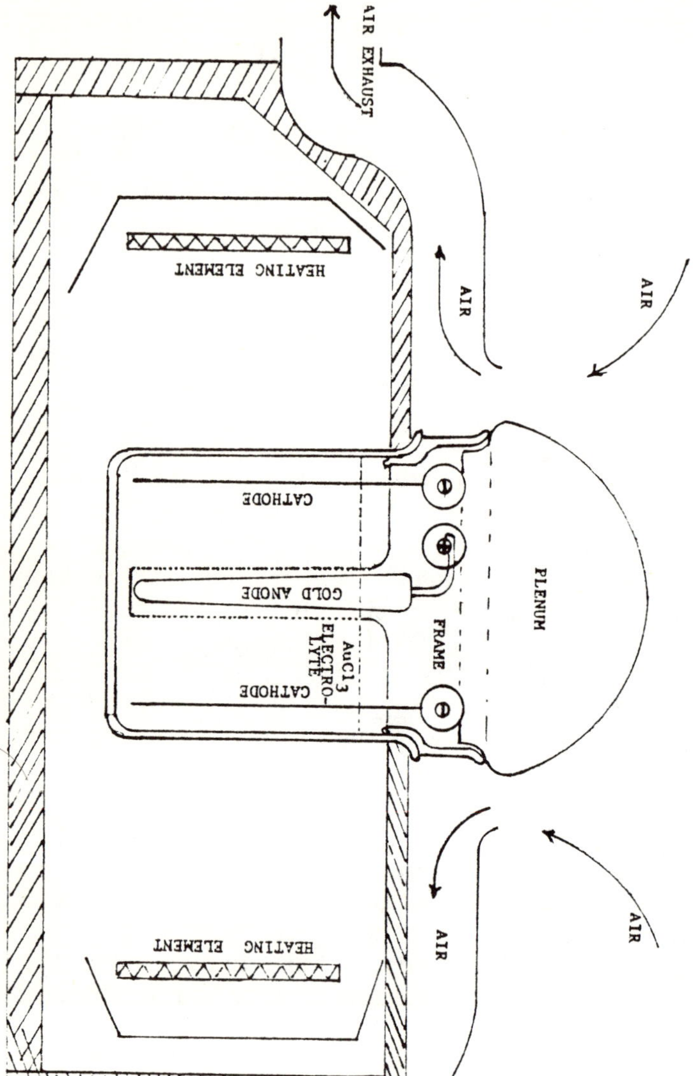

WOHLWILL CELL

ELECTROLYTICAL REFINING PROCESSES.

ELECTROLITICAL CELL	WOHLWILL	MOBIUS	THUM
EFFICIENCY	100%	100%	100%
ELECTROLYTE	75 GRAMS OF GOLD/LITER 10-16% HCl	60 GRAMS OF SILVER/LITER NEUTRAL	60 GRAMS OF SILVER/LITER NEUTRAL
TEMPERATURE °F	158°	69°	69°
CURRENT DENSITY	110-120 AMPS/SQ. FT	50 AMPS/ SQ. FT	50 AMPS/ SQ. FT
CELL VOLTAGE	1.3-1.5 VOLTS	2.7 VOLTS	3.5 VOLTS
RATE OF DEPOSITION	2.452 GRAMS/ AMP. HOUR	4.0245 GRAMS/AMP. HOUR	4.0245 GRAMS/AMP. HOUR
AMPERE HOURS/ TROY OUNCE	12.68	27.82	27.82
OUNCES / AMPERE DAY	1.89	0.86	0.86
KILOWATT HOURS / TROY OUNCE	0.019	0.075	0.09737
OUNCES / KILOWATT DAY	1263.15 (86.62 U.S.Lb)	319.99 (21.94 Lb)	246.48 (16.90 Lb)

glass walls of the plenum. A bakelite cover is placed on the top of the insulated heating mantel so that a very light current of air drawn through the cover is leading away any corrosive vapor which may escape from the cell, expecially when the cell is opened for attendance. This current of air also is providing the necessary cooling of the glass cover (plenum) so that most of the vapor gives off its heat at the glass walls of the plenum.

An electrolytical cell constructed in this manner can operate 24 hours a day without any attendance.

ANALYSIS OF THE PROCESSES

Gold Refining:

Description of any chemical reactions or industrial processes must be based on some assumption of quantities involved, before any consideration can be given to other attributes such as efficiency of the process, economy, speed, etc. All these factors will, of course, affect the final decision in choosing the proper method to reach the objective. It is a general knowledge that in any type of technology or scientific endeavor, there may be more than one valid opinion which is based on sound and realistic consideration. There are always alternative ways to reach the same objective. The feasibility of the whole operation and the economy will finally prevail in making a choice between several types of operations.

The initial description of the stripping process facility was based on the assumption of a yearly gross income of $750,000.00. The actual capacity of the facilities and the calibration of the equipment was adjusted to this size.

The salable product of this operation is primarily gold bullion 9998 fine. The silver bullion, platinum, palladium, limited quantities of metals of the platinum group and base metal scrap may be considered as by-products of the operation.

At an average price of gold on the international market $135.00 per troy ounce, it is necessary to produce within a year 5500 troy ounces of gold to reach the level of the indicated income. The average gross profit of the operation will probably be 25% of the total sales, which is $187,500.00. The cost of the operation and materials will not exceed $25,000.00. The capacity and efficiency of the installation will, of course, affect the number of employees which are necessary to perform the task and also, the final profit of the operation.

Year around operation consists of 250 working days totaling 2000 work hours. Incidental slowing of any part of the operation due to any conceivable cause is not calculated in and is left to be remedied by the management at each of the particular occasions. It is assumed to be feasible to operate chemical processes at 90% of their efficiency. This consideration can be correct if we exclude human behaviour. In any case, we may confine all operations within 1800 work hours per year while calculating all the data at 100% efficiency, except for some instrumentation which may operate without attendance 24 hours a day. It is not advisable to operate any process without attendance over a period of even one hour, however, it is possible, when special quality of the equipment is provided.

The Wohlwill process for refining of gold is just the kind of process where 24 hours a day operation is desirable the reasons are mainly economical. The instrumentation is relatively small and can be designed to the high efficiency and reliability. The electrolyte on the other hand is very expensive escpecially in terms of holding a high concentraion of gold indefinitely in the process, considering that it is necessary to provide another batch of solution for periodical exchange and purification of the foul electrolyte then a considerable amount of gold is involved. A gallon of electrolyte contains 302.6 grams or 9.73 troy ounces of gold in the value of $1,313.71. A larger tank operated 8 hours a day is not as efficient as a small tank operating on the continual basis of 24 hours a day. A constant discontinuation and restarting the process may be time consuming, unless the solution is kept at operating temperature overnight. Then of course, it can stay in operation.

To attain the desired output of gold per year, this process must purify 24.65 troy ounces of gold per day at 90% of its efficiency. This in turn requires approximately 1 troy ounce of gold per hour of operation through the 24 hour working day with 10% of the time down for attendance.

As it can be derived from the table of operational data, the rate of deposition of gold is 2.452 grams per ampere hour. One troy ounce of gold weighs 31.1 grams and therefore, it will require 12.68 ampere hours to deposit the weight of the metal. At the current density at which this process operates (110 amperes per sq. ft.) it will require 16.5999 sq. in. of cathode surface to deposit the weight of gold in one hour. 16.599 sq. in. is equal to 107.09 square centimeters. Using this dimension for a square porcelain tank, we may arrive to approximatley 1 liter of electrolyte which will be necessary for this operation.

To make up 1 liter of electrolyte it will require 80 grams of gold metal in form of auric chloride ($AuCl_3$). The price of a liter of electrolyte is \$350.00. In small quantity of the electrolyte only a very pure "crude metal" can be refined. The operations prior to this process are sufficient to produce an anode pure enough to sustain the operation of this process over a period of 24 hours minimum.

The dimensions of the cast anodes will be approximately 4"x4"x 0.4" (which is 10 x 10 x 1 centimeters). 100 cubic centimetes of gold weighs 1930 grams (62.05 troy ounces). This load of gold will require 62 hours (2 1/2 days) of a continual operation. If an acceleration of the process is essential, then a simple addition of another cathode on the other side of the tank will double the output in weight of purified gold having the anode consumed in one day (24 hours) with some of the scrap anode remaining for recasting.

The starter cathodes are made from a rolled sheet of pure gold.

The foul electrolyte can be stripped of gold by addition of ferrous sulphate and the gold will precipitate on heating. The remainder of the solution can be added to any of the solutions of dental gold or simply the remaining metals of the platinum group in the solution can be reduced by an addition of zinc and then re-processed.

The Thum cell or Moebius cell operation should be maintained on 24 hour basis as in the case of the Wohlwill process. The continuous operation is very feasible due to the low operating temperature (room temperature). The electrolyte, being almost neutral, is not evaporated or decomposed as it is in the case of other processes. For most of the time, it is the water which has to be replenished in order to maintain the necessary concentration and level of the electrolyte. The corrosivity of the vapors to its environment is at minimum. The evaporation can be lowered to some extent by covering the tanks with plastic covers.

The over all efficiency in deposition of the metal is much lower than in Wohlwill process. The cells, if operated at 100% efficiency, can carry only 50 amperes per sq. ft. The rate of deposition of silver is only 4.0245 grams per ampere hour. In comparison to the Wohlwill process it is only 74.59% of the weight deposited in gold, over the same area of the cathodes, within the same time period. The Thum cell cannot be operated at the same efficiency as the Moebius cell, due to the constantly growing barrier of residues building up at the bottom of the basket. It is usually experienced that the cell operates only at 65% efficiency after the initial time, while the Moebius may be maintained at 90%. The current requirements are also much higher due to the higher voltage per cell.

The Wohlwill cell can produce 1263.15 troy ounces of gold per kilowatt day.

The Moebius cell produces 1150.26 troy ounces of silver using the same current.

The Thum cell can produce only 887.26 troy ounces of silver using the same current, and if operated at 100% efficiency, the Thum cell can operate only a short time at this rate. In reality, if these numbers are projected to the expected efficiencies, then the

Wohlwill cell is producing 77.96 U.S. pounds of gold per kilowatt day. The Moebius cell can produce 70.99 U.S. pounds and the Thum process only 35.59 U. S. pounds of siver per the price of processing.

The expense with industrial current supply is not the decisive factor in production of precious metals. It is the size and capacity of equipment which is necessary to obtain the same economical results. Considering the price of gold and silver on the international market, we arrived at a large disparity of efficiency in these processes. Assuming the price of 1968 was $510.37 for U.S. pound of gold and only 61.97 dollars per U.S. pound of silver then the price of silver amounts to 3.14% of the price of gold. In order to reach the same return of profit by processing the silver alone, the process would have to produce 12,102.63 pounds of silver per year. This would require to produce approximately 6.72 pounds per hour and the operation would require 16.82 sq. ft. of cathode. Two Moebius cells would have to operate 24 hours a day, throughout the year.

CHAPTER SIX

PRECIOUS METALS AND THEIR ALLOYS

The precious metals are commercially available in all types of alloys or purities suitable for their ultimate use. A substantial portion of these metals is partially fabricated before being sold. A great portion of the metals are sold alloyed with one or more of alloying constituents to yield the properties required by the user. Pure gold has its own market as a commodity and is generally unaffected by the prices of alloyed gold which serves a special purpose.

Dilution of pure gold with any of the hardeners, creates a departure from the characteristics of the pure metal. Specifically, in strength characteristics, hardness, color, resistance to tarnish and corrosion. Resistance to corrosion in an important factor in the karat golds. Therefore, a certain lower limits are generally accepted as the limits for alloying gold.

The determining factor for production of these various alloys of gold are customs, modified by style trends. For alloys which may be used as a coinage, where obviously something harder than pure gold or pure silver is required, the requirements may assume a technical and legal status.

The state stamping laws and fair trade practices in the United States have defined various precious metal alloys and clad products. The specifications extended over a number of various products concerning the governmental and military supplies, which may originate from many different industries. These specifications in the past have extended even over the type of scrap materials and the legality of processing the materials containing a certain weight of gold per tonnage. Until 1975, it has been determined what processing and purchases must be licensed by the Bureau of Domestic Gold and Silver Operations.

Since 1975, some of these regulations were lifted, however, many of them still prevail. For example, sterling silver must contain not less than 92.5% of pure silver, the remaining 7.5% is unrestricted. Usually, the remaining 7.5% is copper, because copper is an effective hardener.

The gold alloys used in jewelry are described by karats and color. The karat is the content of gold in artilces expressed in 24ths of the total weight. In gold clad stock, the ratio of the weight of the gold alloy cladding to the total weight of the materials, must be indicated. For example, 1/10 12K gold filled is surfaced with a layer or layers of 12K gold alloy that weights 10% of the weight of the composite article. Such an article contains 5% of gold per weight of the article. The designation of "gold filled" is limited by stamping regulations in which the weight of the coating is at least 1/20 of the total weight. This quality mark cannot be applied to articles clad with an alloy less than 10K in quality.

Pure gold can be electro-deposited readily and this method for surfacing is used for high frequency conductors in invironments where silver, which is often used for these purposes, corrodes. A con-

considerable range of gold alloys are also electrodeposited on jewelry and other items where appearance is important. In some instances, several layers of gold and other metals are deposited successfully and then the article is heated to produce an alloy by diffusion of the metals.

Pure gold is resistant to nitric, sulphuric and hydrochloric acids and many other chemicals. However, the use of pure gold in the field of chemistry is very limited. This is due to the susceptibility of the gold to the attack by halogens. The softness and relatively low melting point of gold is another limiting factor for its usage in chemical ware, while platinum and platinum group alloys are generally applied. In some instances, gold is used for surfacing of reflecting mirrors and planes in infrared spectrometers, since the pure gold has high reflectivity in the red and infrared spectral ranges.

In preparation of gold alloys, the furnace which is generally employed, is fed with gas and air under pressure and the melting atmosphere is kept slightly reducing to prevent the oxidation of the base-metal alloying elements. Boric acid or sodium borates are good melting fluxes for most of the alloys. The flux can be skimmed off from the melted surfaces and replaced with fresh flux, just before the metal is poured. Calcium boride is occasionally used for some of the higher melting gold alloys, in particular, dental alloys. To prevent the oxidation of the alloys, the general reliance is placed on 1% or more of zinc in the alloy to maintain a low final content of oxygen. The alloys are poured directly into warm, greased, cast iron split molds which have a long life. During the pouring of the metal and especially when boric acid is not used as a flux, it is desirable to use a reducing flame for covering the metal stream as it enters the mold. The rate of pouring is slow, just equal to the rate of solidification, so that no appreciable shrinking pipe can develop. In centrifugal casting of jewelry items or dental casting, this is not an essential technique.

GOLD JEWELRY ALLOYS

A gold-copper-silver alloy system is the base for all karat yellow gold alloys used in jewelry. Frequently, the properties of these alloys are modified by additions of other elements such as zinc and/or nickel. Zinc is desirable to be present in the alloy to limit the final content of oxygen during the melting operation, but, also for modifying the color of the alloy. The correct proportions between the Cu-Ni-Au-Zn and the required percentage of gold per karat will modify the alloys mechanical and physical properties. Also, the chemical corrosivity and color appearance can be affected. Copper as an alloying element hardens these alloys not only by solid solution effect, but also by low temperature rearrangement ordering of the structure, which causes the age hardening of the alloy. Both copper and silver lower the melting point of the gold alloy. When both of these elements are present, then the solid solubility of copper in silver, and silver in copper, is limited and is decreasing the temperature, thus forming a heterogeneous metal structure which is susceptible to low temperature transformation. The age hardening can be induced in certain alloys by heating the metal for 1 to 2 hours at temperatures between 300-500°F.

THE COLOR OF THE ALLOYS

An addition of 25% of silver to the gold will result in an 18K alloy which has a good green color. By replacing about half of the silver with copper will produce a yellow gold alloy, and the further increase in copper will produce a red gold. In a 14K alloy (58.3% of gold) 41.65% of silver produces an unattractive greenish-white color and also the metal is too soft to be useful. If 6-12% of the silver content is replaced by copper the green color may somewhat improve. A presence of zinc in such an alloy will add to the appearance of greenish color. If 20-30% of copper is present, then the yellow colored alloys are formed. A further increase in copper in the alloy yields reddish metal. The presence of little nickel (1-3%) will lighten the color and also harden the alloy. Zinc will also lighten the color of the alloy but will add slight greenish cast and softening of the alloy.

Useful compositions are confined to the higher copper content of thc alloy with a considerable amount of silver added which will yield gold alloys with good yellow coloration. Very often, zinc and nickel are added to improve the physical properties of the alloy. It seems to be very desirable to produce an alloy which may have an appearance of pure gold and the hardness of brass. The appearance of pure gold may not be as desirable as one would estimate at first glance, as a matter of fact, most jewelry doesn't have the appearance of pure gold. The gold in jewelry is expected to have an intensive, bright yellow color and hardness which is not natural to pure gold. It is also expected a certain workable strength in the alloy. As a matter of fact, the customers usually don't know how pure 1000 fine gold really looks. Yellow gold color is due to the fact that the major portion of the reflectivity of the gold is in the visible range of the spectrum and a great portion of the reflectivity is in the infrared regions of the spectrum. An addition of silver to gold apparently reduces the reflectivity in the extreme red, however, if added to the alloy in excess it will also reduce part of the reflectivity in the yellow portion of the spectrum. The reduction of reflectivity is the reason why the silver ultimately produces green coloration instead of white coloration which would originate from increased reflectivity across the visible spectral ranges. Therefore, the addition of silver has to be limited only to reduction of reflectivity in the red regions and thus producing the naturally yellow color of gold. This could be achieved in very high karat golds. The limited addition of silver in lower karat golds would not be sufficient to make up the necessary difference in weight in 14 and 18K golds. 41% of the weight must be added to the gold to produce 14K alloy. The limits of addition of silver is set to 19% of the base metals added, this means that a 14K gold has 58.35% of gold and the remainin 41.65% of the total weight of the alloy must be made up from other materials. 39% of that weight can be substituted by silver, this represents for the total weight of the alloy approximately 16-17% of silver. The rest of the weight can be filled with copper which may contain 1-2% of zinc and approximately 1-3% of nickel. This addition produces a good yellow appearance of the gold which does not reflect any light in infrared.

The same yellow color can be achieved in 18K gold where gold is present in 75% of the total weight and the remainder can be made up with the same proportions of the alloy. The additions of the hardener and the de-oxidizer which is the nickel-zinc, if kept constant through the variation in composition of the karat gold, will ultimately affect the color appearance of the very high alloys of gold, such as 22K gold. Due to the fact that the addition of 2-4% of this metal to the total weight of the alloy will start to get more prominent in the composition of higher karat golds. A 22K gold which composition is 91.5% of pure gold and 2-3% of hardeners with remaining 6% of the base alloys will assume a light yellow color. Unlike it is believed, a 22K gold alloy can be produced by this addition, which is hard and suitable for manufacturing of jewelry and can be polished to a high luster. A constitutuion of the gold karat alloy system is presented graphically in the diagrams.

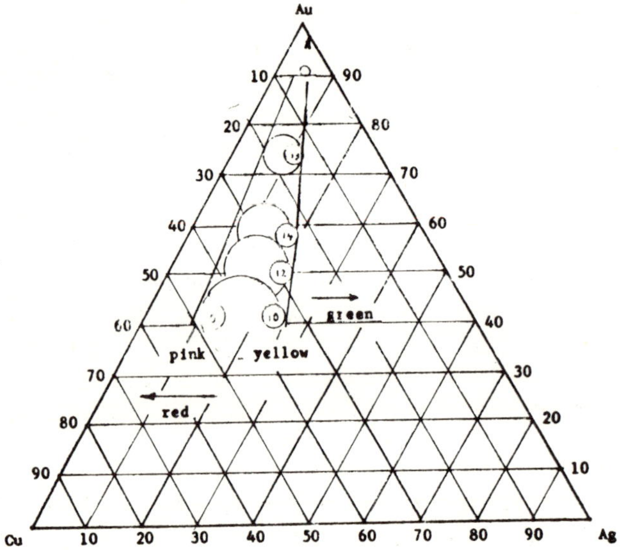

Approximate composition of gold alloys that make relatively yellow and pink metals of various carats.

TERNARY DIAGRAMS

For readers who are not acquainted with graphical presentation of chemical composition of an alloy, especially the ternary diagrams, some word of explanation is necessary.

Ternary diagrams usually represent a composition of three elements in the alloy system and the effect of this chemical composition on its physical properties. The physical properties which were determined by testing in the laboratories are usually plotted in some graphical presentation that they represent the relationship of this sought property to the varying compositions of the alloy and that one may discern and project the trends of expected values. This can, of course, be done in a very simple way for binary alloys where the total mass of the chemical composition consists of two elements and by varying one's percentage the other element must be the reciprocal value. These two elements are plotted in the horizontal abcissa where one element is increasing in percentage toward the end of the graph while the other element is at the same time decreasing in its percentage. The physical properties such as hardness or others can be plotted on the vertical ordinate, and in this manner the hardness is related to the varying composition of these two elements by a curve. If the graph is extended to other elements substituting one of the elements in the original composition (the composition still remains a binary alloy) then an additional curve can represent this relationship on the same graph.

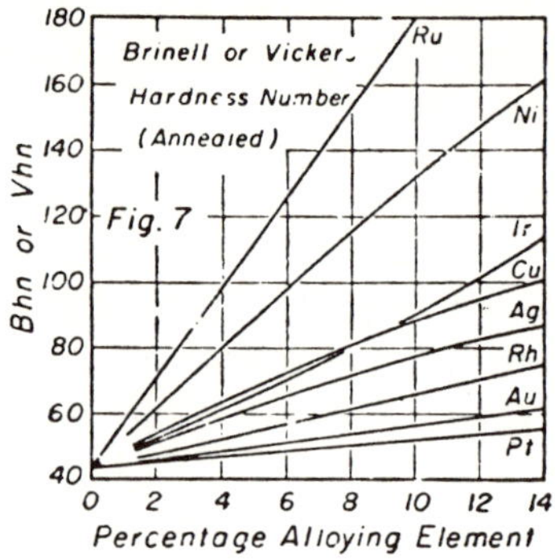

If an alloy system has more than two elements present, then the composition of the major three elements which decidedly affects the physical properties of the alloy is plotted on ternary diagrams. Sometimes, only the base metals added are plotted on a separate diagram assuming the base metal to be always present in its proportion to the total.

Each element in the composition is plotted on its own percentage scale which is a side of an equilateral triangle. Each percentage scale starts from 0 to 100% in the next corner of the triangle, therefore, each corner represents 100% of the element in the composition, and total absence of the other in the composition of the alloy. In turn each corner of the triangle represents 100% of the element in the composition. Any point on the percentage scale represents the percentage of that particular element with remaining of the percentage of the adjacent scale of another element. That means binary alloy. Any point within the plane of the triangle represents the composition of all three elements in the proportions as they can be derived from the ordinates. The center of the triangle represents a composition of equal portions of all three elements.

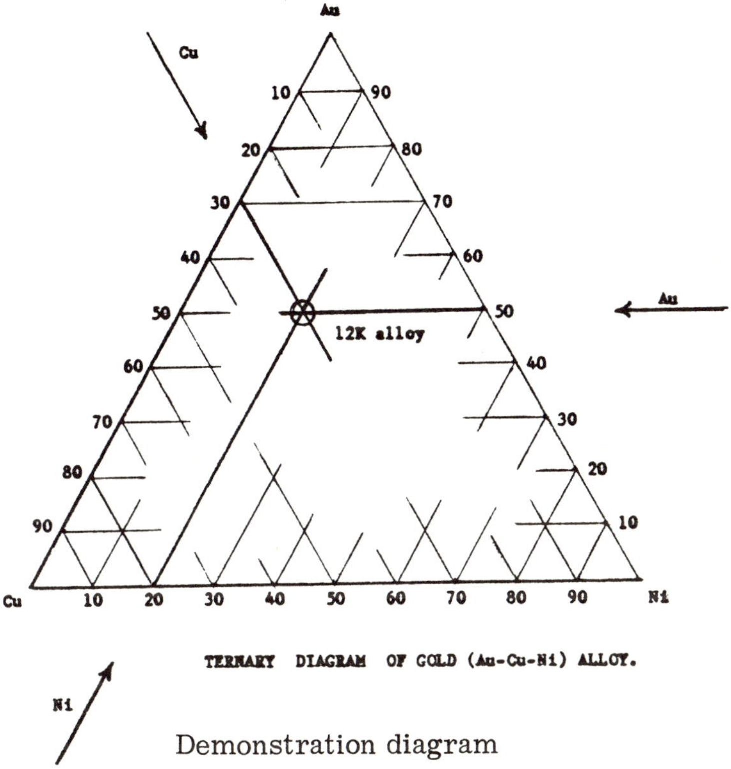

Demonstration diagram

Each data on the diagram represents a tested physical property of various compositions of the alloy. Connecting the points of the same readings of the hardness for various compositions will produce a three dimensional value superimposed on the diagram. This numerical reading of the tested data is a fourth scale in the diagram, located on the coincident compositions of the alloy and therefore representing the relationship of hardness (physical property) to the composition of the material.

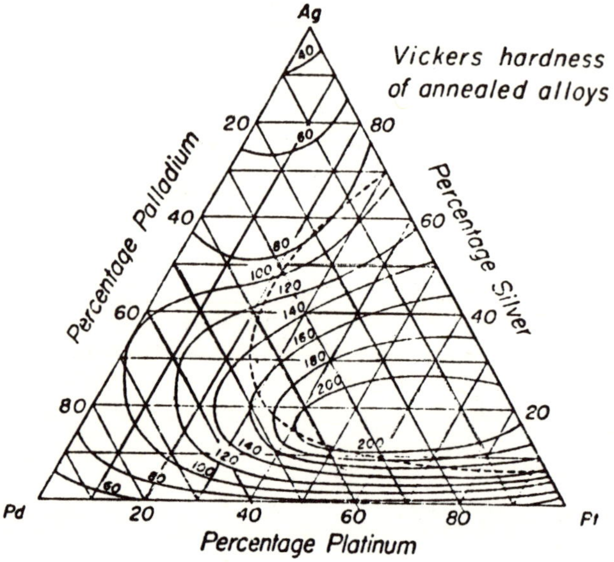

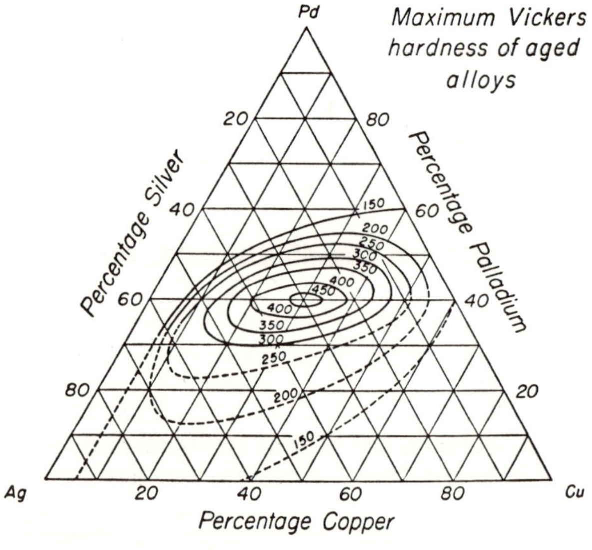

PINK OR ROSE GOLDS

Only certain alloys of this series are used in jewelry manufacturing. For most of the time, it is applied to frames for eyeglasses. These alloys are used in solid form or are clad on a core of silver, nickel or bronze. The pink golds are by no means stainless or tarnish resistant. They darken by perspiration or by exposure in urban atmosphere. In general, these alloys present difficulties in manufacturing due to the susceptibility to corrosion, and stress corrosion cracking. Some alloys which are free from zinc seem to be somewhat less susceptible to corrosion cracking. Also, marketing of this type of materials for ornamental purposes is rather low. Most applications of this type of alloys can be seen in optical instruments as reflectors for infrared spectra.

Compositions of Commercial Pink Gold Alloys

Composition, %					Density, g per cu cm	Solidus, °C	Yield Strength, psi
Au	Ag	Cu	Ni	Zn			
							14-carat
58.3	4.9	31.6	5.2	...	13.08	949	48,000[b]
58.3	3.3	31.0	3.5	3.9	12.37	938	43,800[b]
							12-carat
50.0	4.0	37.1	4.3	4.6	12.17	954	......
50.0	....	42.0	6.0	2.0	12.19	999	......
							10-carat
[a]41.6	9.2	41.0	1.9	6.3	.....	...	......
41.6	4.6	43.4	5.0	5.4	11.60	952	38,700[b]
41.6	6.8	44.3	7.3	...	11.55	980	54,100[b]
41.5	11.7	46.8	...	...	11.2	...	41,500[c]
41.5	....	51.5	3.7	3.3	.....	...	......

WHITE GOLD

The white gold alloys originally were offered as a substitute for platinum in jewelry. These alloys originally contained 80% of gold, later 18K, 14K and 10K alloys were developed in the same manner as yellow gold alloys. The hardness of these alloys is greater than of the yellow gold alloys. Palladium has been very grequently used instead of nickel. The content of platinum or palladium in white golds places them in another category than jewelry. For most of the time, golds containing platinum and palladium are classified as dental alloys. An 18K gold-nickel alloy quenched from 1560°F. is a homogenous and hard alloy somewhat ductile, white in color and resistant to nitric acid. If the same composition is cooled down slowly, the same alloy becomes heterogeneous, yellowish white in color and susceptible to the attack by nitric acid. This already by itself indicates that this alloy undergoes solid state low temperature transformation which particularly affects the appearance and physical property of this alloy.

The composition ranges as they are given in the following graphs are results of a compromise between the color and working properties of this alloy. The composition limits are rather critical, especially in the higher karat alloys.

The highest karat alloy in the white series, which can be produced is 19.4K alloy. Any further increase in percentage of gold will affect the coloration of the metal. This alloy is usually improved in ductility by addition of 1 or 1.5% of copper. This addition reduces the incidence of edge cracking during the rolling operations or forging. This addition does not seem to change the color of the alloy too much.

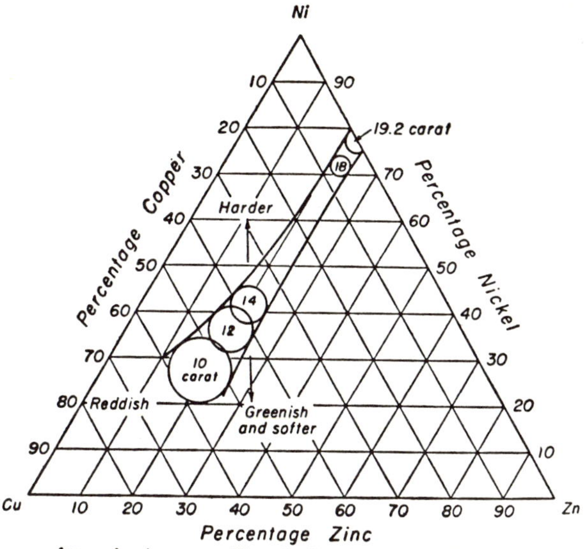

Approximate composition limits of the base-metal alloy that is added to gold to make relatively white alloys of various carats

White gold ternary diagram

As it can be noticed in the diagram there is a whole series of lower karat white golds. The composition of the base metals to be alloyed with an appropriate amount of gold is shown in the diagram.

The limitations in the use of these alloys is mainly due to cracking during the severe reductions in stamping and rolling operations. The cracking develops during the heating of stressed materials. Slow heating is not alleviating this problem. It is necessary to minimize the stresses in parts before annealing.

In general, white golds resist tarnish better than colored golds of the same karat, but the resistance of white alloys to general corrosion is probably less because the atomic percentage of gold in white alloys is lower for a given karat.

DENTAL GOLD ALLOYS

The American Dental Association provided specifications for several types of dental gold. Also, federal specifications have been established for several classes. Many of the commercial dental alloys fit into this classification, however, many dental alloys have been designed to meet special commercial requirements. The most widely used alloys in dentistry are shown in the following tables.

Composition Limits of Cast Gold Dental Alloys

No.	Type	Composition, percentage Au	Ag	Cu	Pd	Pt	Zn
				Yellow Golds			
1	Soft	79–92.5	3–12	2–4.5	0–0.5	0–0.5	0–0.5
2	Medium	75–78	12–14.5	7–10	1–4	0–1	0.5
3	Hard	62-78	8–26	8–11	2–4	0–3	1
4	Partial denture...	60–71.5	4.5–20	11–16	0–5	0–8.5	1–2
				White Golds			
5	Hard	65–70	7–12	6–10	10–12	0–4	1–2
6	Partial denture[a].	60–65	10–15	9–12	6–10	4–8	1–2
7	Partial denture[b].	28–30	25–30	20–25	15–20	3–7	0.5–1.5

(a) Yellowish white (b) Very white

For demonstration purposes some charts are included for variety of products which are on the market.

Constitution of Binary Alloys

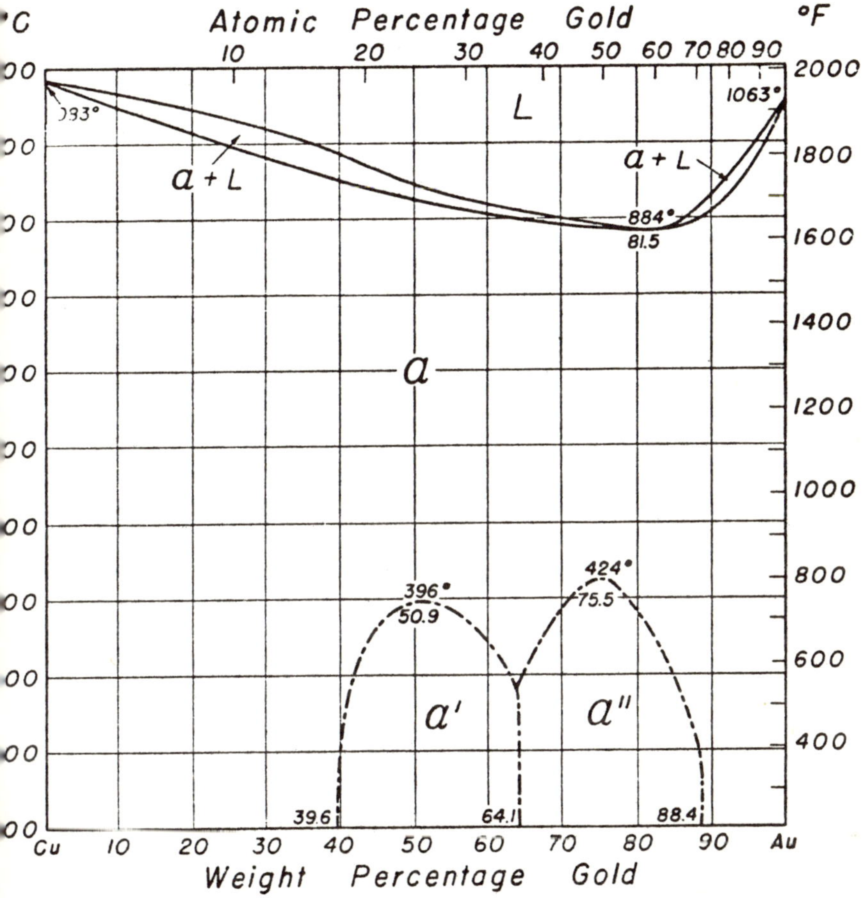

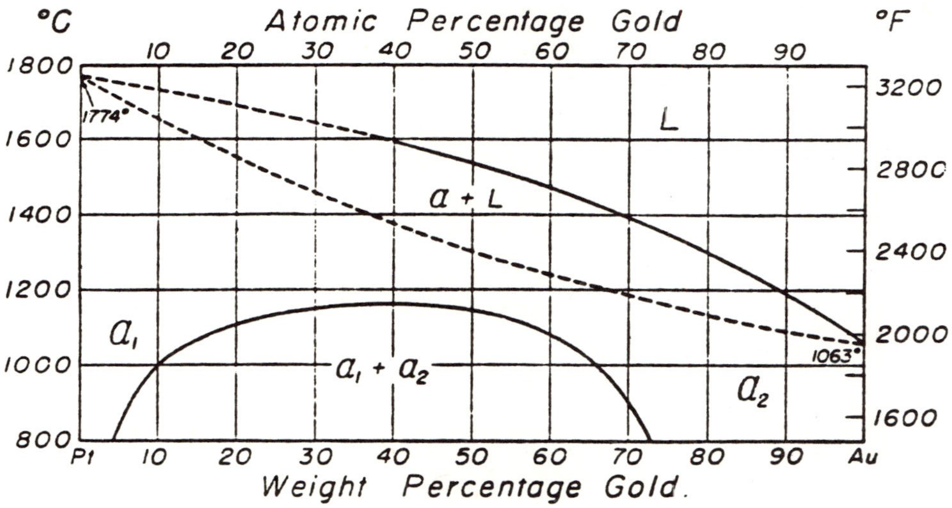

Au-Pt Gold-Platinum

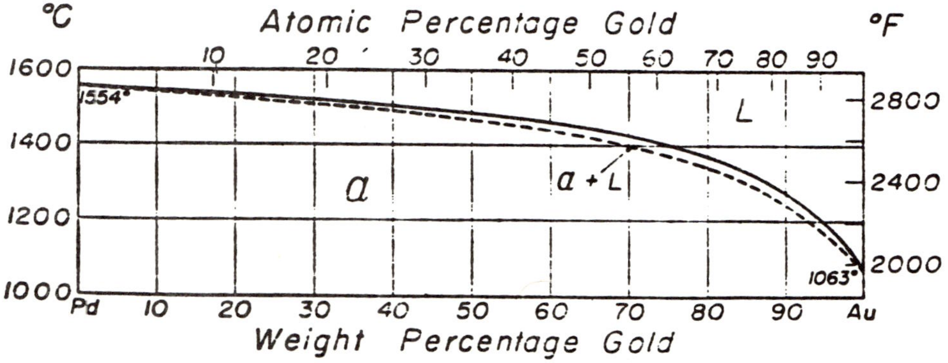

Au-Pd Gold-Palladium

SILVER AND SILVER ALLOYS

Oxygen dissolves in pure molten silver. Some of the oxygen is released again when the metal solidifies. Silver does not oxidize by contact with the air, however, the oxygen content should be brought to a low level before the metal is poured. This is very readily accomplished by having graphite or powdered carbon on the surface of the metal during the melting. The function of graphite is not to reduce the silver oxides to metal but to prevent the access of air to the metal surface by reacting the carbon with the oxygen and thus, creating the protective atmoshpere during the melting. This protective layer serves also the purpose of protecting the iron crucible, which is used many times for melting the silver, to be excessively oxidized.

Where a high electrical conductivity is required, purity must be maintained, as in the case of copper, but iron need not be guarded against. The solid solubility of oxygen in silver and the rate of diffusion of oxygen through the hot metal are very high, consequently, silver of high purity will be damaged if it is heated under oxidized conditions. High purity of silver is desirable for its high heat and electrical conductivity which exceeds the properties of copper. Silver offers advantages over copper in application of electrical contacts because of its resistance to oxidation. Other industrial uses are based on corrosion resistance, the catalytic activity and also, the ease of electro-deposition of the metal.

In braising the alloys, the silver alloys have moderately low melting points and good mechanical properties. The high and uniform reflectability in visible regions of the spectrum is the reason for usage of silver on mirrors, silverware and jewelry.

The use of silver alloys for coinage is a custom thousands of years old. The reasons were probably the appearance of the metal and its relative scarcity.

The photo-sensitivity of silver halides and the ease with which silver can be reduced are the basis of photography.

The density of pure silver is 10.49 at 20°C. Sometimes the density is lowered by cold work or high absorption of oxygen.

The melting point for oxygen-free silver is 1760.9°F. (960.6°C.)

The rate of volatilization is quite high at elevated temperatures. Silver metal has partial pressure of 100 milliliters of mercury at 1870°C. A prolonged heating of silver prior to pouring will result in loss of the metal even when alloyed with other metals. Silver completely volatilizes at 2212°C. after it reaches its boiling point at 760 milliliters of mercury pressure.

SILVER-COPPER ALLOYS

The copper is an effective hardener for silver but lowers its electrical and thermal conductivity. The silver-copper alloys have been used and are still used for coinage silver. 10% copper and 90% silver alloy is called "coin silver". This alloy can be hardened by aging. The coin silver as well as sterling silver will respond to an aging treatment of 2 hours at 280°C. or 1 hour at 300° C. or 30 minutes at 325°C. Higher copper alloys containing 20-30% of copper do not age harden at the temperatures for the silver rich alloys. The hardness is usually lower after aging.

STERLING SILVER

Sterling silver must contain at least 92.5% of silver. The remainder is unrestricted but it is normally copper. Other metals have been proved to be less effective hardeners. The hardening of sterling silver is the same as of coinage silver or other copper-silver alloys. It can be done in protective atmosphere at 650°C. and then the metal can be quenched to room temperature with subsequent artificial aging at 325°C. for 30 minutes.

BRAISING ALLOYS
SILVER-COPPER-INC (Cadmium-Tin)

These silver alloys are used principally for joining metals and are known as silver solders or silver braising alloys. The quaternary alloys have varied percentages of cadmium and tin in addition to copper and zinc. These alloys have much lower liquideous temperatures. Since the diffusion of the braising alloys into the joint members varies with the type of metal or alloys being joined and the time and temperature of heating, the selection of proper variation of alloys is important with respect to each specific job. The mechanical properties of the joint will vary with the type of joint and composition of the material being joined.

SILVER FOR ELECTRICAL CONTACTS

Un alloyed silver can be used for electrical contacts. Under higher current conditions certain silver alloys are preferred for greater serviceability. Addition of alloying materials to silver will increase the electrical resistivity. The metals usually added to the silver to form alloys for electrical contacts are tungsten, molybdenum, nickel, and sometimes substances such as cadmium oxide, lead oxide, and even graphite. The methods of powdered metallurgy, rather than melting, are used for compounding of these alloys. (SINTERING)

DENTAL AMALGAM

The dental amalgam has 33-52% of silver, 12.5% of mercury, 2% of copper and 0.5% of zinc. The typical uses of this amalgam are to restore the lost tooth structure, sometimes known as silver fillings.

In processing of silver this amalgam must not be loaded into either one of the processing tanks since the mercury would severely contaminate the electrolyte and destroy the electrolytical processes. Also, in analytical determination of silver percentage in amalgam the thiocynate volume metric method can not be used. The silver content in amalgams must be determined by precipitating the silver as AgCl out of the solution as it is described in method 5B. The amalgam is dissolved in nitric acid and filtered to remove the stannic acid which will form on boiling in concentrated nitric acid. The filtered solution is diluted with distilled water to 3 times its volume and then the silver can be precipitated by addition of sodium chloride. The silver chloride can be collected by filtration. The silver can be reduced to metal in fusion with sodium hydroxide and then purificd in Moebius cell.

PLATINUM AND PLATINUM ALLOYS

Platinum is the most known element from the entire platinum group. The physical properties of platinum places this metal mainly for industrial uses, especially where special techniques are employed and harsh chemicals are used for fusions which otherwise would destroy ceramics and glass. The ductility of the platinum enables the technicians to apply the metal for various purposes in the laboratory. The high conductivity, high melting point and general corrosion resistance are very desireable properties for industrial uses.

Chemically platinum is an outstanding catalyst, both for oxidation and for hydrogenation in production of vitamins and other chemicals. Platinum-iridium catalysts are widely used in petro-chemistry.

Platinum is used for electrical contacts, chemical production, laboratory equipment, anodes in both solid and clad form, jewelry, and for decorative purposes.

In addition, various alloying elements such as ruthenium, iridium and for special purposes other hardeners are used to develop higher mechanical properties. There are four general classifications or grades of wrought platinum which contain at least 99% of the element. Commercially, platinum is classified as follows:

TYPE A: Type A platinum is 99.99% pure and is used primarily for its physical properties. This grade is sometimes called "physically pure" because the evaluation of the purity was based on the thermal emf (electro-motive force) and temperature coefficient of the resistivity. These physical properties were determined in addition to the spectro-chemical analysis. This type of material is used in thermocouples and resistance thermometers.

TYPE B: Type B contains usually 99.9% platinum and is called "chemically pure". This metal is used for preparations of chemicals such as platinum salts or for making up the alloys for catalysts for instance, platinum-iridium or platinum-rhodium catalyst, etc.

TYPE C: The "crucible grade" contains 99.5% platinum, and is used for general purposes such as anodes in electrolytical processes. The high corrosion resistance of this metal, its high inertness, the conductivity and high melting point, are the features which are very desirable and can be obtained from this grade of material.

TYPE D: This type is known as " commercial platinum". It contains 99% platinum. This type of material is used for electrical contacts, electrical breakers and also in jewelry. The resistance to corrosion of all these grades of platinum at ordinary temperature is the same, however, at very high temperatures under some extreme oxidizing conditions the differences can be observed.

Platinum should not be used in chemical reactions or heating in presence of elements that will diffuse into the metal. Many of the base metals may permanently contaminate the platinum surfaces when heated to the diffusion temperature. Some of these physical properties are indicated in the table for the four types of commercially available platinum alloys.

Some Physical and Mechanical Properties of Commercial Platinum

Property	Types A	B	C	D
Electrical resistivity, microhm-cm				
0 C	9.83	.....	.....	.....
20 C	10.6	10.8	11.4	14.9
Temperature coefficient of electrical resistivity				
0 to 100 C	0.003923	.....	.....	.....
20 to 100 C	0.00364	0.0036	0.0035	0.0034
Thermal emf (hot end 1200 C, cold end 0 C) versus high-purity Pt such as Nat Bur Standards Pt 27, millivolts	0 to plus 0.030 mv	to approx plus 0.150 mv	to 1.0 mv	.....
Tensile strength (annealed), 1000 psi	17 to 19	20 to 23	24	26
VHN (annealed)	38 to 42	42 to 46	50	52
Tensile strength (hard drawn), 1000 psi	34	35	35.4	45.5
VHN (hard drawn)	97	100	110	130
Annealing temperature, 50% reduction	650 C	.....	.....	.....
Annealing temperature, 75% reduction	550 C	750 C	850 C	950 C
Hot working temperature	800 to 1200 C	800 to 1200 C	800 to 1200 C	800 to 1200 C

Table of the four types of platinum.

PLATINUM RHODIUM ALLOYS

Rhodium as an alloying material can be found in several types of alloys ranging from 3.5% up to 20% of rhodium in the alloy. Rhodium is a preferred addition to platinum in most applications. Such as reactions at high temperatures under strong oxidizing conditions. Rhodium is a hardener and it is not selectively volatilized from the alloy as it is the case with other alloying materials. Platinum-rhodium 10% alloy is the most important of the series. It finds its use as a catalyst for the oxidation of ammonia by air or in production of nitric acid. The same alloy with very precisely controlled composition serves as a positive element in the standard thermo-couple which uses the electrodes platinum-rhodium versus platinum. This thermo-couple is widely used for high temperature control industrial instruments.

10-20% alloys are used for winding of high temperature resistance furnaces which operate under oxidizing conditions. The temperature resistance and the corrosion resistance of these alloys is better than the ones of pure platinum metal.

PLATINUM IRIDIUM ALLOYS

These alloys find their use in equipment for chemical processing and to some extent in dentistry. The 25% iridium-platinum alloy is used for electrical contacts, hypodermic needles, and other articles in which maximum hardness plus ductility at high annealing temperatures is essential. For high temperature service in air these alloys are not used since iridium can be volatilized and be lost from the alloy by oxidation.

PLATINUM RUTHENIUM ALLOYS

The 5% ruthenium-platinum alloy, same as 10% iridium-platinum alloy, are used in jewelry and are known as "hard platinum" of the jewelry trade. Of course, these alloys are also used for electrical contacts in airplane spark plugs and for printed resistors in electronic microcircuitries.

84Pt-10Pd-6Ru Alloy

Combination of platinum, palladium and ruthenium or some other combinations of platinum with less noble elements such as nickel or tungsten found a place on the market. The properties of these alloys serve a specific purpose for which they were designed and therefore, they are not as common as the above mentioned alloys.

PALLADIUM AND PALLADIUM ALLOYS

Palladium resembles platinum in many respects, in appearance, ductility and strength. Palladium is in abundance and its price is approximately 25% of the price of platinum with which metal it is closely associated. The uses of palladium is very similar to that of platinum. The primary use of palladium is in dental alloys, in jewelry, and as an industrial catalyst in the manufacture of synthetic organic products such as vitamins, ethylene, and also in the manufacture and purification of hydrogen. Hydrogen can be purified by rapid and selective diffusion of hydrogen through palladium metal and in this manner a chemically pure hydrogen can be produced. Sometimes the effect of hydrogen affinity to palladium is used to act as a valve that will admit hydrogen to vacuum systems when the palladium is warmed.

60Pd-40Ag

This alloy is used for electrical contacts of low temperature coefficient.

60Pd-40Cu

This alloy is also used for electrical contacts in the milliampere range in circuits, containing sufficient capacity, so that a considerable rush of current occurs on closure. It is also used where a hard material is required for slip rings running against brushies of the same material. This material requires heat treatment to convert it into ordered and highly conducting condition.

PALLADIUM-SILVER-COPPER ALLOYS

The variations of this alloy are used in dentistry. The alloy contains 35% palladium, with sufficient copper that it can be age hardened.

PALLADIUM-GOLD-SILVER ALLOYS

Palladium-gold-silver alloys and palladium-platinum-silver alloys are the variants of dental alloys.

GLOSSARY OF TERMS

Alloy - a material composed of two or more metallic elements, intimately mixed usually being fused and dissolved in each other when molten. At elevated temperatures, the alloying metals may form a solid solution from which new intermetallic compounds may precipitate on cooling, thus forming the hetergenous structure of metals.

Anneal - a heat treatment process at specific temperature at which an alloyed material is re-crystallized. It is usually accompanied by a change of physical properties of the metal.

Angstrom unit Å - a unit of length equal to one ten-thousandth of a micron or $1x10^{-10}$
These units are used to express wave lengths of light

Anode - positive electrode in electrolytic cell. A sink for electrons.

Aqua Regia - a mixture of nitric and hydrochloric acid. This mixture is capable of dissolving gold.

Archimedes principle - the principle states that an object sugmerged or partly immersed in liquid looses its apparent weight by the amount equal to the weight of the liquid displaced.

Auric - is an oxidized ion of gold to trivalent state.

Aurous - name pertaining to gold compounds in which gold is univalent.

Binary alloy- an alloy consisting of two metals only. See alloy.

Carat - a measure of proportion of gold in jewelry alloys. One carat means one part of gold in 24 parts of total metal by weight, or total of 100 divided by 24 is equal to 4.16% per weight.

Carat - also a unit of weight for precious stones, equivalent to 200 milligrams.

Cathode - a negative electrode in electrolytic cell.

Cell - see electrolytic cell.

Cementation - a displacement of one element by another from the solution. The displacing element is higher in electromotive series than the element being displaced.

Coulomb - a measure of electric charge; a charge transferred by one ampere in one second.

Cupel - a small, shallow porous cup used in assaying gold. It serves the purpose of separating the precious metals from lead which in turn absorbs other contaminating elements in the sample.

Cupellation - a process of assaying in which gold and silver is placed in a cupel and subjected to oxidizing fusion with lead. The molten oxides are absorbed by the porous cupel leaving the gold-silver button on the surface. See scorification.

Cyanoauric acid - a colorless crystalline acid $HAu(CN)_4 \cdot 3H_2O$, which decomposes on heating into gold and hydrocyanic acid HCN.

Dental gold - is gold alloy containing silver, copper and some of platinum group metals such as palladium.

Denver crucible - a clay crucible of parabolic or

nearly parabolic cross section useful for torch melting of materials or precious metals. The flame played on the surface of the metal is deflected and will emerge away from the torch tip.

Dore bullion - unparted gold and silver in bars.

Electrolysis - separation of chemical compounds into their constituent parts by means of electric current.

Electrolyte - a solution of chemical substances and/or metal ions which is a strong electric conductor. The current is carried by ions and matter is liberated at the electrodes. See Electrolytic cell.

Electrolytic cell - a vessel containing electrodes, anode and cathode, and an electrolyte. Passage of an electric current is accompanied by liberation of matter at the electrodes. The amount of material liberated at the anode or cathode is in proportion to the electric charge passed (Faraday's Law); one Faraday of charge (96,500 Coulombs) will cause liberation of one gram equivalent of substance (its atomic weight divided by its valence measured in grams.)

Electrum - a naturally occuring alloy of gold and silver.

Fineness - the purity of the metal generally expressed as part per thousand.

Free milling ore - ore in which gold or silver is in natent form and easily amalgamated or leached by cyanide.

Fulminating gold - an explosive substance obtained

by the action of ammonium hydroxide on gold chloride.

Gooch crucible - a small crucible with perforated bottom in which the precipitate can be collected by filtration, dried and weighed.

Gravimetric analysis - chemical analysis in which amounts of constitutents are chemically separated and their weight determined.

Halogen - an element of the chlorine family; chlorine, bromine, iodine and fluorine.

Kt - abbreviation for carat. See carat.

Liquidus - physical state of matter most of the substances can exist in three basic physical states, solid, liquid or gas. The relationship of this physical state toward the environmental conditions under which the transition may occur is usually expressed in form of liquidus curve indicating the melting point, boiling point or freeing point in the system.

Litharge - lead monoxide in fused form as opposed to massicot which is lead monoxide in unfused form.

Micron - a unit of length equal to one millionth of a meter, a micro meter.

Miller process - a process for refining gold in which chlorine gas is bubbled through the molten metal. Base metals are converted to chlorides which are volatized. Silver chloride which is also formed remains in molten state and is taken up by flux or partially volatized. The remaining gold is 995 and requires additional processing. This method is not suitable to remove any contamination by platinum metals.

Moebius cell - or Moebius process is an electrolytic parting process for silver.

Oxidation - (1) a chemical reaction with oxygen (2) a chemical reaction which results in a metallic element moving to a higher valent state.
(3) loss of an electron.

Parting process - separating of two metals of the same family by extraction or electrolysis.

Perodic table - also Mendelyeev's table, a table of elements arranged in a systematic way and grouped in periodically occuring groups according to their atomic number. Elements in the group tend to have similar chemical and sometimes physical properties.

Platinum black - is a metallic platinum in form of fine black powder obtained by reduction of the metal from its solution usually by addition of formic acid or others.

Platinum sponge - metallic platinum in form of grey porous and particulate sponge usually obtained by thermal decomposition of ammonium chloroplatinate at 1000°F.

Policing - collecting or guttering precipitates adhering to side of vessel.

Proof gold - pure gold 1000 fine.

Refractory ore - ore in which metal occurs. Minerals which are difficult to treat; the opposite of free milling ore.

Recovery - amount of metal obtained in a process of ore treatment, sometimes expressed as a percentage of metal in the ore. It expresses the yield which may vary by methods applied. It does not represent the absolute percentage of the element in the ore.

Reduction - (1) chemical reaction in which oxygen is removed from the substance. (2) chemical reaction which results in a metallic element moving to a lower state or zero valence state - metallic state. (3) gain of an electron.

Scorification - process in assaying in which gold and silver bearing substance is mixed with lead, borax and other flux in a crucible and is subjected to oxidizing fusion to produce slag (scoria) containing impurities the gold and silver is retained in a lead button. In scorification volatile materials form a slag as opposed to cuppelation where the same materials would be absorbed by the cupel.

Solidus state - see liquidus.

Specific gravity - ratio of weight of a volume of substance to the weight of an equal volume of water. Specific gravity of gold is 19.3 and 10.5 for silver as cast.

Spectrometric analysis - a method of chemical analysis whereby the concentration of an element is approximated to the intensity of the light emitted by the atoms of the sought element at a specific wave length which is specifically related to the kind of atoms of the sought element. The spectrum is excited by external electrical sources.

Sterling - (1) a measure of fineness. Originally fineness of lawful coin of England; for silver 995 fine (old sterling) today, 9166 fine. (2) fineness of 925 for tableware or other manufactured silver articles, the remaining 075 being copper added for hardening.

Supernatant - liquid above the solidus matter usually precipitated from the liquid.

Titration - method in chemical analysis in which a measured volume of liquid reagent is added to the known volume or weight of sample in liquidus state. The volume is determined by adding the reagent until some definite effect is observed such as color change, or turbidity. For this reason special indicators are added to the sample solution. Upon reaching the final point the liquid in the buret is accuragely measured and the volumetric proportions are taken in calculation of the final result.

Tow tray - tray used for transporting batches of materials through the manufacturing installations.

Volumetric methods - method of chemical depending upon use of definite volumes of standard reagents. (See Titrations)

Wohwill process - an electrolytic process for refining gold using super-imposed alternate current over the DC current in a chloride electrolyte.

Zone refining - a refining technique in which a difference in melting point of contaminated metal and pure matter is used as method of transport of the contaminates from the mass of purified metal. As the molten zone is moved along at rated speed, the pure material is freeing out while lower melting contaminated material is advancing in the direction of the furnace. Several passes (in one direction only) are necessary to render the material pure.